108

ALBUQUERQUE ACADEMY

SCIENTIA SIT POTENTIA

1955

This book was donated
to honor
the birthday of

Elise M. Sena

by

Richard Sena

2007

DATE

ALSO BY BRUCE STUTZ

Natural Lives, Modern Times: People and Places of the Delaware River

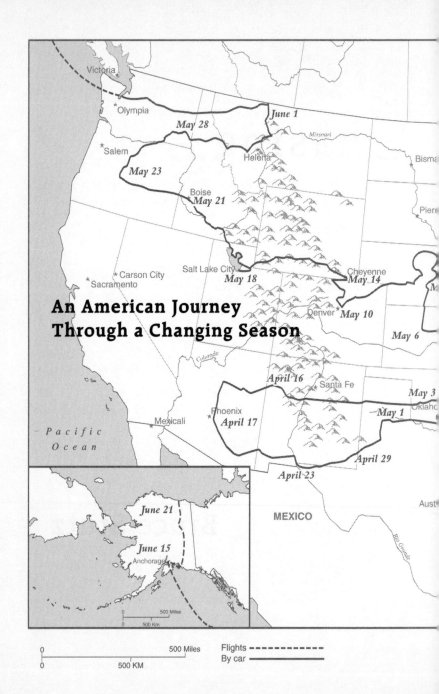

An American Journey
Through a Changing Season

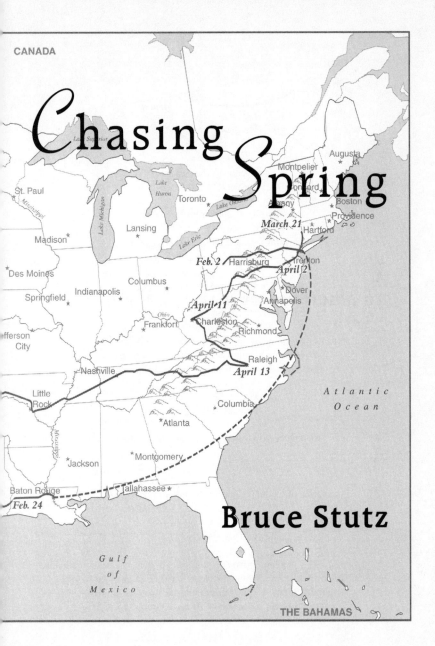

Chasing Spring

Bruce Stutz

Scribner

NEW YORK   LONDON   TORONTO   SYDNEY

SCRIBNER
1230 Avenue of the Americas
New York, NY 10020

For information about special discounts for bulk purchases,
please contact Simon & Schuster Special Sales:
1-800-456-6798 or business@simonandschuster.com

Designed by Kyoko Watanabe
Text set in Bembo

Manufactured in the United States of America

1   3   5   7   9   10   8   6   4   2

Library of Congress Control Number: 2005052905

ISBN-13: 978-0-7432-6247-7
ISBN-10: 0-7432-6247-6

For Benjamin, Nathaniel, and Julia

Contents

Chasing Spring

The Pumphead in Winter

November 21
9 HOURS, 43 MINUTES OF DAYLIGHT

When setting out on a journey I always take note of the light. In this way, I orient myself and track my course. And so from my hospital bed this morning, awaiting surgery to repair a busted heart valve, I look out at the dawning wintry sky through windows streaked by New York City soot and note that though my view extends across the East River and its phalanx of still-lighted bridges, I see no sign of the sun.

I wonder what this gloom might portend, perhaps because some primal part of me believes light has meaning—that sunlight brings clarity and hope, that an overcast sky invites vagueness and disappointment. Or I might just be feeling the effects of whatever drugs they gave me a couple of hours ago when, while still dark, they wheeled me to an examining room, stood me up under harsh lights, threaded a tube from my groin to my heart, and through it streamed a dye that searched for arterial blockages. I had to be awake, the doctors explained from behind their blue masks, and I understood this to mean there was nothing to be done for the searing pain that accompanied the tubing's every vermiculation. If they found nothing, that was good, and they would then operate.

For most of my life, I've been aware of my murmuring heart. Whenever I am alone, in the car, at my desk, lying in bed, I can feel it. I can count its beats until I fall asleep. I've even timed my hiking to its rhythm. Sometimes my heart gulps as if it is trying to swallow something too large. Sometimes it leaps into double time. I never thought this was unusual. I thought everyone could hear his or her heart.

Back in my room now, I'm waiting. The pain from the inserted tube is already unrecallable. But I'm wondering how much more is to come. I'm wondering if I'll live. "I have to tell you," my surgeon put it, "that there's a one-out-of-a-hundred chance you might die during the surgery." I had not yet signed on the dotted line, had sat in his office only long enough to see the photos of his family, his sailboat, and the framed crayoned thank-you notes from young patients. I nodded, tight-lipped, as I'd seen it done in the movies, and I signed. Not out of fearlessness: the fact was that if I didn't have the operation I could well end up with inoperable congestive heart failure. The odds that I'll die today, you can look it up, lie somewhere between those of coming up with a straight and those of a full house, both of which, in playing very few hands of poker, I've been dealt.

So the dull dawn is beginning to get to me and I find myself thinking what I'll do come spring and realize this is what we all think when stricken by wintry doubts. We think, *Come spring . . .* and mean that we're ready for change and that we'll get to it in the season when everything around us begins changing, when vagueness fades and we can proceed with purpose, one with the greening shoots, swelling buds, and lengthening days, attuned to Earth's durable rhythms of light and time. Even those of us whose lives might be barely touched by nature the rest of the year will, come spring, be seduced into a hazy synesthesia wherein smells beget hopes and light induces longings.

How does it happen? In my hospital bed I close my eyes and recall my springs of the past, remember the first day when it was suddenly warm enough to ride my bicycle to school or take out my baseball glove (oiled, folded over a hardball in its pocket, and stored all winter with a rubber band around it) and could quit flipping cards across the bedroom floor while it snowed outside. I remember the first top-down days for my yellow 1966 Corvair and warm nights making out in it while fully and painfully clothed. In college, marijuana wafted through the air on good spring days. I remember meeting my wife-to-be in a spring rainstorm that we walked through in blissful ignorance, our clothing saturated with the perfume of the season's first blossoms. On the first warm days with the kids on the playground, she and I would sit with the rest of the parents on the sun-soaked park benches. My first spring hike always took me up Kittatinny Ridge to the rocky overlook above the Delaware Water Gap where the low-bush blueberries and the mountain laurel had just begun to bud. Along the river below, as the spring waters receded, the banks burst forth with tangled vines of wild roses and strawberries.

The last week in April my wife, Sallie, the kids, and I would head out into the just greening woods of New Jersey to pick morels, mushrooms whose appearance meant spring had arrived. We'd take the short hike to our own spot, a little slope by a waterfall that poured its spring runoff down the faces of gleaming shales. The dog by now knew the way and trotted on ahead while we walked with our straw baskets and our bag lunches along the path whose early spring blooms we knew as well as those in our own garden. Here the violets, here the imperious swamp cabbage, its stalk wrapped like Balzac, here the curled fiddleheads of ostrich ferns, and here the still tightfisted mayapples. All waiting to unfurl.

We'd cross a little stream that percolated out of the ground and that seemed to run each spring with the same unhurried burble. Songbirds had not yet appeared so the forest had a secretive silence but for the stream and the tattling waterfall a hundred yards up ahead that misted the adjacent forested slope enough to spawn the little crop of morels. The waters' sound invigorated the old dog and she hied up into a trot. The kids would run too, for we'd promise a dollar for the first morel.

The morel's a most feted fungus. Its phallic shape, firm feel, and earthy taste imply, and not subtly, fertility. Its appearance in spring is celebrated in mountain forests across the country. While most mushrooms appear in the late summer and fall when there's more for them to grow on, morels take advantage of spring's first sweet stirrings, sucking sugars from the growing trees' first photosynthesis and in turn providing the trees with digestible nutrients. This fungus is more ancient than the plants whose flowers and fruits are easy pickings. And yet in a couple of weeks they're gone, to appear ever farther north until the spring finally ends. If unpicked, they'll dry up or in spring rains deliquesce and the spores that line their fibrous caps will be dispersed on the wind.

By the time we reached the summit of the little slope, the dog was already shaking off the water from her swim. We'd begin to search in the forest duff beneath the elms for a sighting. We knew they were there, but morels are always there and yet not there. They don't show themselves. Camouflaged among the fallen leaves and twigs, there could be fifty and you wouldn't see one. But their smell, musky and languorous, wafts up with the mist from the falls.

Finally, to see them, we'd lie down, all of us on our bellies, flat out on the ground, and turn our heads in the hope that by scanning the dark palette of the forest duff from this vantage, we'd see, against our low horizon, morel mushrooms standing. If at this moment a hiker came through the woods, I don't know what he'd think. I don't know what we'd tell him. Anything, I guess, but that morels grow here. I remember realizing that the mushrooms were not quite as important as the ritual. Of the going out in the woods in spring. Of our children pressing their cheeks to moist warming soil and knowing the exhalations of the spring earth.

And then they'd appear, seeming to arise magically before our eyes.

"I see one!"

"I see one!"

"There's another!"

"Dad, don't move, there's one right by your foot!"

We'd gather what we could, place them into the wicker basket, and then sit scanning the ground for any we might have missed. Only after our attention left the ground did we notice the sun through the canopy of elms and once again the sound of the meltwater pouring through the limestone karsts of the hillside and spilling out from the rocks—the purging of winter.

Bring on that juice and joy! I'm ready now for the spring forest in its mist, spring in its musk, in its spring greens: canopy green, shafts of wild iris green, understory green, diatom green in slow running stream, buoyant sun-dappled lily pad green, and rime-varnished watercress green.

I am all the more ready for spring, I confess, because I'm feeling that my most verdant years are behind me. Little love, big love, lost love, people lost, time passing. I knew my great-grandparents, immigrants who spoke very little English and so talked to me, their beloved first American great-grandchild, with such a wild vocabulary of gestures, chucks, pinches, and grins that I was more frightened than comprehending. I enjoyed the lavish love of grandparents and the fretful love of my parents. And then I watched my father suffer long years and my mother tender him ceaseless attention. He died only a year ago, six months before my diagnosis, and the last time I saw him—I knew it would be the last time—he lay beneath the hospital sheets, and his large frame seemed to have no mass at all. His hawk brown eyes worked to remain keen. I imagined a grueling palaver going on behind them. Disabled by successive surgeries, he spent most of his last years doing battle with his body. Everyone admired his tenaciousness, mistaking it for a will to live. I was convinced that he lived for the love of his grandchildren, in whom, I realized, he recognized the transformative energy possessed by growing, blossoming, transmuting things. They were of him, the wellspring of his persevering as well as his offspring.

I wonder if my own kids, when they came to see me last night, thought me as helpless as I thought my father. Seeing one's parents weakened or humiliated must be a young child's most difficult moment. It's not so easy for a grown child, either. So I tried to let them know I was up for this. And if not, if I die in surgery, then I've told my children many times how much I love them and how much I trust—no, more—admire their minds and decent sensate hearts.

We can't be children in just-spring again. But being intimate with such energy is every bit as good as being a part of it. And so we're affected when spring comes with its spawning, nesting, breeding, brooding, rising lignin, peeping peepers, its skunk cabbage and mayapple, crocus and magnolia bloom, dandelion and daffodil, when spring comes on the wind and osprey call from familiar hangs, when shorebirds flash across the marsh at dusk, when songbirds sweep into forests to chirrup May and butterflies flutter on solitary flights, when all, all flock determinedly north: fishes up ancestral streams, salmon, shad, quicksilver schools of herring and big-eyed hag-faced lamprey. We sow crops, play ball, leave our jackets on the playground at recess, be free and in shirtsleeves, for then comes Hinamatsuri, Holi, Passover, Easter, Rowanfest. Then comes light, resurrection, renewal, joy of survival from heart surgery, joy of spring when, chorused Thomas Nashe, "blooms each thing . . . the pretty birds do sing, cuckoo, jug, jug, pu-we, to-witta-woo!" and when, effused Gerard Manley Hopkins, "weeds, in wheels, shoot long and lovely and lush"; and when, D. H. Lawrence wondered, "what fountain of fire am I among / This leaping combustion . . . ?" Then we want to be a part of it all.

Better come quick, for I hear the gurney rattling down the hall. "Let's go," says the masked man.

Between the quick hands of the attendants I catch glimpses of my wife, my mother, and my younger son. The bed rails collapse with the clacking of an ironing board being closed. Someone gives me an injection. I see lights pass as if I'm half asleep in a cab passing through the city after midnight. I'm wheeled into a cold room with a stainless steel ceiling.

In this room the surgeons take my life from me by chilling my body temperature to 77 degrees and injecting a cold potassium solution that stops my heart for a two-hour break after 53 years of beating its offbeat rhythm. While the heart-lung machine pumps cooled blood through my hypothermic body, the surgeons do their work. This is not your old-fashioned hacksaw-the-sternum, pry-open-the-rib-cage surgery but a refined postmillennial procedure accomplished with nanocameras and instruments, which went into the heart through an incision made beneath my right breast, that the surgeon manipulates from outside my body, like an unseen God-like puppeteer deftly working his marionettes.

All presumably goes well because I wake in a dimly lit room where a nurse sits at a desk reading the newspaper. I try to speak to her but can't get out a word and at once feel the pain of a breathing tube that's been shoved down my throat. I begin to gag, my eyes tearing, and gesture toward the nurse. When she looks up I point to the tube and let her know that I want it *out*.

She pays patient attention to my frantic hand-waving, turns her wrist to look at her watch, and says, "You need to have it in for another hour or so."

It hurts when I shake my head but I do so emphatically and begin a choking fit, at first forced but finally very real. She puts down her paper and walks over to me.

"I'll ask them to come and take it out."

I nod as well as I can and lie back, trying not to swallow. She walks off into the dark where I assume other patients lie, zombies waiting to awaken.

Well, I'm alive, I think, when suddenly, the patient in the next bed bolts upright as if a spring had snapped in his spine.

"Not here!" he shouts. "Not here! They're going to kill me here! Take me to the animal medical center!"

The nurse returns, turns on the light over the man, whose wide eyes stare out blankly from a pale and anhydrous face, and presses him slowly back down to his mattress.

"I've been here before," he protests weakly. "They're trying to kill me here."

On second thought, maybe I'm not alive at all, just lying in Hell's waiting room, my first torture being this tube in my craw. In fact, other than this, I feel nothing at all. And then I notice that I don't hear my heart. It's gone. I listen hard. Nothing. Either I've died or I've gone from stem-wind to quartz movement. I'm amazed that time can advance in silence. This will take some getting used to.

I decide I'm alive. Now what?

A week later I leave the hospital and return to my house in Brooklyn with a sack full of medications. Because the doctors didn't saw me in two I'm actually out and about the very next day, perambulating the block on the kind of slow, shuffling stroll old Mafia dons take in their slippers. The first few days when I return from my once-around I barely have the energy to make it up the five steps of the front stoop. When I get inside I can't make it up the stairs to my bed and just fall asleep on the sofa.

I'd figured this would go on for a while but nearly a month later when my walks are longer and my step surer, I still feel hazy— disconnected and lost. The surgeon's warning that "You may be a little depressed for a time" doesn't do justice to the no-man's-land in which I find, or can't find, myself. Reading up on the possible causes of post-op existential shock, I discover in *The New England Journal of Medicine* a report that found my symptoms among patients connected to a heart-lung machine during surgery.

The researcher called it "cognitive decline." Heart surgeons, he said, had another name for it—"pumphead"—a name so apt it's never uttered in the presence of patients. The syndrome is caused by microdebris or microbubbles released during heart surgery, small enough to pass through even the finest of synthetic cloth filters but large enough to cause tiny embolisms in the brain's capillaries. So, along with having lost my heartbeat, I'm also suffering synaptic disjunctions from rerouted blood. I wander, a pumphead in winter, left to notice on each of my brief sojourns that each day there are fewer minutes of daylight, that each morning the sun passes a bit lower to the horizon, and that the shadows that shadow me have shadows of their own. I begin to realize why our ancestors feared winter's failing light.

With so much time on my hands and feeling a need to put winter into historical perspective, I take up Sir James Frazer's monumental *The Golden Bough,* in which the 19th-century Scottish classicist and anthropologist prospects cultures' myths, rituals, and fables in search of shared connections. No paradigm seemed to Frazer more overarching than that of spring's rebirth.

Could the "primitive savage . . . be certain that the luminary would ever retrace his heavenly road?" he asks. "It was natural, therefore, that with such thoughts and fears he should have done all that in him lay to bring back the faded blossom to the bough, to swing the low sun of winter up to his old place in the summer sky . . ."

Throughout December I walk on a primitive's watch. As darkness comes earlier more holiday lights of Christmas and Chanukah appear. The world is still inspired by ancient rituals devised in the hope that the right magic, the right fire at the right time, will affect the course of natural events and spur on the solstice. That day, December 21—the shortest day of the year—the transition from winter into spring begins.

Soon, whole blocks stand festooned. In the week before Christmas, carolers stroll from house to house with lit tapers. In my own home we begin the eight-day celebration of Chanukah that commemorates the rededication of the ancient Temple in Jerusalem. In this spirit each night we add another candle to the eight-armed menorah so that, like the sun after the winter solstice, each day it will burn a little brighter.

Christmas finds its sun in the actual birth of a god whose light the Wise Men followed to his birthplace. In the nativities of Correggio, Murillo, and Caravaggio, light infusing the dawn at the birth of Christ dawns all the brighter on the face of the newborn. (Lighted crèches on front lawns re-create the same scene. One I see features life-size plastic figures illuminated from within. They're all there: Jesus, Mary, Joseph, Rudolph, Mickey, and Goofy.)

In the Julian calendar, the solar calendar devised in the first century at the request of Julius Caesar, the winter solstice fell on December 25. This happened to coincide with the day the Syrians and Egyptians celebrated the nativity of the Sun, when at midnight celebrants cried out, "The Virgin has brought forth! The light is waxing!"

That the Church placed the birth of Jesus on the day that pagans celebrated the return of the sun is no coincidence. They did it, writes Frazer, "in order to transfer the devotion of the heathen from the Sun to him who was called the Sun of Righteousness."

Christ, like all the other gods that secured the light in midwinter—Tammuz, Attis, Adonis—will die young. These gods must die so that they don't live through the year's corruption and decay—the wintry exigencies of life—that we mortals must experience. And so they might be resurrected again come spring.

In my shuffling, shadowed state, I wonder whether resurrection might be more than metaphor. My grandfather was the only person I knew who claimed to have been resurrected, and not just once. When I was about nine years old we lived next door to my father's parents, and one morning my grandmother called to say she thought my grandfather was having a heart attack. My mother rushed over but, in her panic, didn't realize that I followed right behind her. At my grandparents' bedroom door I saw my grand-father lying lifeless on his bed, his head up against the wood head-board with its carved swirls, the great tasseled shades of the bedside lamps casting a parchment glow no brighter than candles. My grandmother was wailing, "Lou! Lou!" My mother stood holding her. Dr. Katz, a solid ball of a man with a dapper mustache, had arrived on the scene, thrown off his sport coat, climbed up on the bed, put his knees on my grandfather's chest, and, like a cowboy on a bull, pumped up and down until the ambulance arrived.

When my mother realized that I'd witnessed it all she took me to the hospital to visit my grandfather to show me that he had lived. At that time children weren't welcomed into hospitals and I remember my father explaining to the nurse why I was there. I wasn't certain myself whether I wanted to see him or not but I went in. He lay beneath a plastic oxygen tent, his hands by his sides. He waved me closer.

"Nurse," he said hoarsely, "this is my grandson. If you're look-ing for a young man to take to dinner one night . . ."

"No talking," she said.

"How are you feeling?" I asked him.

"How? I'm feeling very rejected."

"No talking," the nurse said again.

"What do you mean, 'rejected'?" I whispered.

"They told me my heart stopped three times," he whispered back. "I was dead. I think somebody up there doesn't want me." And he smiled and brought his hand out from beneath the plastic tent, his hand with its heavy veins and coarse hairs. And he patted my hand and closed his eyes.

Knowing now what it's like to hover, pumpheaded, between past and future, I wonder whether, when my grandfather touched my hand, he did it for my sake or his own. He lived a long time, suf-fered a few more deaths and rejections, and every Easter reminded everyone that Jesus wasn't the only Jew ever resurrected.

So why not me?

December 21
9 HOURS, 15 MINUTES OF DAYLIGHT

I decide to mark the solstice by taking my first journey beyond the immediate perimeter of my home and head uptown to meet my friend Neil Tyson at the American Museum of Natural History, where he is director of the Hayden Planetarium. Neil is a Princeton physicist, writer, magazine columnist, and oenophile. He is also an ebullient and irrepressible explainer, the kind of person one calls upon when metaphors just aren't doing it. His sources are cosmological and so about as primary as one can find. As with many scientists I've worked with over the years, he has a particular world in mind to which he can refer both to what he knows and what he doesn't know, the grand as well as the microscopic.

For instance, driving in the car with Neil and talking about the observation of space he stops midsentence to assert, "Anyone who doesn't wear a seat belt has no understanding of physics." As we get stuck behind a bus he lets me know that if I had a better grasp of the acceleration of moving bodies I would have been able to pass. When I can't hear out of my cell phone he explains that due to the way the circuits work I'd hear better by putting my finger over the voice slot than by putting my hand over my ear.

This is the kind of knowledge I want. Neil and I go across the street from the museum for lunch and I tell him that I'm interested in the nature of spring. I say nothing more in expectation of an expository torrent.

He thinks, and then says, "The change of seasons is due to the singular fact that Earth, in its orbit around the Sun, is tilted 23 degrees or so relative to the Sun."

He stops there because the waiter's arrived to take our orders. But afterward he says nothing, his very large round brown eyes staring, with some boyish satisfaction, into mine.

"That's it?" I ask.

"That's it," he says.

"I don't understand," I say.

Now he's ready. I can tell.

"As Earth circles the Sun, its orientation—the direction of its tilt—never changes: The North Pole always points toward the Pole Star, Polaris, and the South Pole always toward the constellation Octans. The hemisphere tilted toward the Sun experiences summer while the hemisphere tilted away experiences winter."

I've written this down but I experience the same nausea that gripped me back in my ninth-grade earth-science classroom when I asked how it was possible to see the Sun and the Moon in the sky at the same time. I'm hoping for a grown-up epiphany but recall only the many science fair re-creations of flashlight Sun, soccer ball Earth, and Ping-Pong ball Moon suspended in wire hanger orbits.

"In the summer hemisphere the Sun strikes more directly . . ."

Neil raises himself from his chair. He's a big man with long arms and large hands, so when he stands in for the Sun he also becomes the cynosure of the lunch crowd.

"From the solstice the hours of daylight increase until, a quarter orbit away, there are 12 hours of daylight and 12 hours of darkness. This is the spring equinox, March 21, midway between the winter and summer solstices. Another quarter orbit away . . ."

His arms outstretched, he turns on his substantial axis, a quarter orbit at a time. The waiter, bringing our platters, has to side-step Neil's celestial equator.

I am writing like mad but I'm just not seeing it. Maybe it is still my pump-addled brain. But Neil is making his point clear: Spring is the result of a singular astronomical alignment that rules the chemistry of the planet, the behavior of its plants and animals, and cultures and customs of its people.

While we eat he fills in the details: "Because of that 23-degree tilt the times of sunrise and sunset depend on your latitude. The closer you are to the poles the more extreme the difference between summer and winter light. Deep into the Arctic winter you'll never see the light of day. For weeks during the Arctic summer, darkness may never come at all."

This, I realize, is what I want to experience, a beginning, middle, and end—9, 10, 12, 24 hours of sun, from south to north, from equinox to solstice—a narrative of light and warmth. I imagine surfing a wave of green as it sweeps north, catching the first bloom; first birds, first thaw, first rains, first flood, first leaves, first fruits, plowed fields, desert flora to tundra bloom, as Gerard Manley Hopkins put it, "Growth in everything—/Flesh and fleece, fur and feather,/Grass and green world all together."

"You'll have to begin before the equinox," Tyson says. "It's the date we set for the beginning of spring but daylight begins increasing now and by February it will already be having its effects."

I will have to begin right away. Spring is coming.

January 1
9 HOURS, 19 MINUTES OF DAYLIGHT

January arrives, a new year. I arrange a shelf full of guidebooks to states, parks, birds, trees, wildflowers, geology, and astronomy, and I spread out road maps, topographic maps, state maps, migration routes, and National Park and Forest maps and begin to plan an itinerary. In February and March, depending on how I'm feeling, I'll have to make a few individual forays. If they go well, come the equinox, I'll head south through the Appalachians, then west across the southern plains to the Arizona desert, and then north through the grasslands, into the Rockies, to Montana, and then to Alaska. My goal is the Arctic the day of the solstice—June 21—when, on the last day of spring, daylight lasts a full 24 hours.

Spring, I know, is not a single sweep of warming, greening, migrating from south to north. Each place, each species, experiences its own spring—the valleys before the mountains, south-facing slopes before those slopes that face north. Birds may depend on the changing hours of light to begin their migrations, but each bird has its own timing mechanism as well as its own navigation system. Insects hatch when the temperature allows. A cold hard rain can keep migrating fish from moving upstream.

Spring is the sum of all these individual responses to the changing light, warming air, earth, and water. Small changes can bring few or countless transformations. And each year is different. Even if we thought we might catalog every stimulus, account for every force, the fact is that nature has no obligation to respond to them in the same way. The unexpected is to be expected but can't be counted on. And who could count them? By the time of the Vernal Equinox, spring in the northern hemisphere is advancing north at a maddening pace.

But something else is happening, too. Over the winter as I collect items—from newspapers, scientific journals, information from researchers I know—I find gathering evidence that spring has been changing, that across the northern hemisphere spring is arriving earlier, as much as a week earlier, in some places more, than it did as little as 30 years ago. The reports come from every quarter and lend a growing sense of immediacy to my journey.

Spring thaw comes sooner. In the mountains snow melts earlier. Arctic sea ice is vanishing. Biologists as well as amateur watchers spot frogs, birds, bees, and butterflies arriving before their usual times. Insects that appear before the plants flower are left with no pollen on which to feed. If insects disappear before the birds arrive, the birds, in need of energy for breeding and nesting after long flights, go hungry. Should frogs spawn too early in spring their eggs stand the chance of freezing in a late frost. The season bears down on most of these animals with evolutionary urgency. Spring is the one time of year they have available to breed.

The concatenations of the changes are only beginning to be seen. In Europe, the winter moth that times its hatching to the sprouting of young oak leaves has begun to hatch weeks before the oaks leaf. Their decline will leave arriving birds in the lurch. The retreat of Arctic sea ice has left seals and walrus with fewer platforms from which to hunt fish and mussels. The Yupik Indians report the walruses they hunt have become thinner and their blubber less nutritious. Killer whales that hunt walrus have begun to hunt sea otters. Fewer sea otters will mean more sea urchins. More sea urchins will devastate the ocean kelp. Less kelp will mean less habitat for fish that are the main source of food for whales and the Yupik.

These ecological rondeaux are no longer playing offstage. We can recognize them in our gardens and backyards, and scientists who once regarded as subjective and inconsequential yours and my vague notions of change—"I can recall when . . ." and each of us can fill in the rest—now acknowledge that our primitive sense that there's something up may be based on more than anecdote or nostalgia for lost farmland, forest, and fresh water.

In fact, the seasonal paradigm is changing. Anyone who gardens knows it. Since 1960, the U.S. National Arboretum along with the American Horticultural Society has published a map of climate zones to give gardeners a key to which plants are best suited to local weather conditions. Since their main concern was with landscape plants that must survive the winter outdoors, they divided the country into 10 zones based on 10-degree differences in the average lowest winter temperatures. Up in Fairbanks, Alaska, for instance, Zone 1, where winter temperatures can drop to 50 below, the landscape choices run from dwarf birch to quaking aspen and Lapland rhododendron. Florida gardeners, in Zone 10, can be certain to have success with bougainvillea and palms. Gardeners in the temperate zones between or on the borders of one zone and another can choose to plant sure bets or push their luck. The irregular corrugations of the zones were based on average temperatures from the 1940s to the 1960s.

The map was changed in 1990 because from the 1960s on, the climate belts began to broaden and the borders between them blur. In 2003, a new map made it clear that a northward zonal drift was in progress. Chicago and Denver, which had been in Zone 5, in which average minimum winter temperatures ranged from 10 to 20 below, now found themselves in Zone 6, in which the lows were 10 degrees warmer. Zone 7, on the 1960 map a pink belt that runs only as far north as the Maryland border, in the new map extends into southeastern Pennsylvania and across New Jersey. Gardeners there now have their chance to try landscaping with English yew and holly, less hardy hybrid azaleas, and they might put their young flowers and tomato plants out earlier than they did a few years ago. The downside is that those plants and bulbs that need cold winters in which to lie dormant may no longer grow where they once did.

A 30-year study of 100 species of plants in the Washington, D.C., area by the Smithsonian Institution found that 89 of them now flower an average of four and a half days earlier. The capital's cherry trees blossom a full week earlier so tourists have to plan accordingly and soon, I imagine, local celebrations of spring across the country to mark the harvests of wild ramps and mushrooms, and the blossoming of everything from cherry trees to azaleas, dogwoods, magnolias, lilac, and dandelion, will have to be rescheduled to account for climate change.

Since 1950, the year I was born, the world's annual carbon emissions have gone from 1.5 billion to 6.8 billion tons. The carbon dioxide level in the atmosphere has increased from 310 to just under 380 parts per million—and done so at a rate faster than any time in the last 20,000 years and is accelerating, last year by nearly twice the annual average of the previous 50 years. Earth's average temperature has risen from 56.8 to 58.2 degrees Fahrenheit. The world's population has gone from 2.5 to 6.5 billion. Over my lifetime I've seen corridors of suburban growth crowd out the open fields and forests along the east coast from Florida to Boston. Metropolises have grown up in the western deserts. Heart surgery may have, for the moment, predisposed me to nostalgia, but even among my son's friends, who are in their 20s, I hear, "I remember when . . ." for the sprawl has not diminished.

It all boomed for the benefit of me and my generation and, since most of us never experienced cataclysms such as the Depression or world war, we could choose to imagine that catastrophes didn't happen any longer, or if they did they would come with a mushroom cloud, with such suddenness and force that we wouldn't be able to do anything about it anyway. Now we have a new century, a wireless century, in which things can communicate without being connected. We can log on when and where we want. We can do surgery without touching the patient. However we define this era, it ought to be making us realize the urgency of preserving the nature of the planet if only for its being a bastion of correlations and sensuality. I'm reading naturalist and writer Edwin Way Teale's journey "North with the Spring," written the year I was born and in which he tracks bird migrations from Florida to Maine. Following along on a map and in my mind, I hardly recognize the landscapes or the culture through which he travels.

I'll be seeing a very altered spring.

February 2
10 HOURS, 9 MINUTES OF DAYLIGHT

The pumpheadedness dissipates in this way: I think it's gone and I realize just how confused I was. In a couple of weeks I realize that I'm clearer still and have to face the fact that I haven't been as clear as I had thought. It's much like when the optometrist flips his series of lenses in front of your eyes. But I'm feeling better, stronger, and decide to make a first spring reconnaissance, like the groundhog, to see what it's like out in the world.

Every year when I was growing up in Pennsylvania, a few days before February 2, the little hamlet of Punxsutawney at the western end of the state seemed to suddenly appear, like a Brigadoon or Camelot, and for a few days the state's newspapers, radio, and television focused on whether Phil, the oracular groundhog, would predict an early spring or consign the northern hemisphere to another six weeks of winter. It wasn't until I left the state that I realized how much the greater world focuses on Phil, who, if he didn't live year-round in a comfortable cage in the Punxsutawney Public Library eating dog food and the occasional ice cream cone, might be a fitter oracle. His record is dismal. Over the last 100 years, a throw of the dice each February 2 would have given better results.

Cynics, however, are not welcome in the dead of winter in Pennsylvania. Those who are going to mutter under their breath when housebound know enough to exile themselves to the nearest frozen lake where they can sit on an upturned grout bucket, fishing rod in hand, and stew in the cold while staring into a hole in the ice. For the rest, the annual groundhog pageant comes as a welcome winter break (this was pre–Super Bowl spectacular) and with willing suspension of disbelief we granted Phil the title "Seer of Seers, Sage of Sages, Prognosticator of Prognosticators, and Weather Prophet Extraordinaire." Why forgo a ritual that, as Frazer puts it, preserves "something of the flavour and freshness of the olden time, some breath of the springtime of the world"? So I pack my necessary meds, set the sun at my back, and drive west from Brooklyn to experience the festivities and take in what feels to me, despite the meteorological evidence to the contrary, like the first breath of spring.

Any hopes I had for an early thaw are quickly dashed by my drive through the rolling foothills of the Alleghenies. I'm met by surly skies, crystalline snowfields, and bare forest choirs. Despite the cold, I keep my window rolled down, for my senses are famished. The road follows a river's icy winding course. On a frozen lake, a lone figure sits fishing. Christmas lights still dangle from the rain gutters of vinyl-sided homes; the smell of wood smoke wafts on the wind. I encounter a cold mist, a steel bridge, a stone ale-house, and then, Punxsutawney.

The little town brims with traffic. Festivities have commenced. The Pantall Hotel, built in 1888 on the town square, remains the only hotel and, with its three stories, the tallest building in town. All of its 75 rooms have been booked solid for months, leaving me and thousands more to try and find rooms at roadside motels. I spot a vacancy 15 miles from town, check in, and return to Punxsutawney.

A chilling mist has set in but, in the park across from the Pantall, dollar beers and charcoal-smoked kielbasa warm up the crowd. Bundled in parkas, the pilgrims ruminate among rows of makeshift kiosks that display local arts and crafts from marmalades to crocheted berets. But the groundhog rules. The groundhog is the feted oracle. No pagan spirit could have asked for more veneration; no ancient animist could have rendered more devotion. I see grown men got up in groundhog anoraks with furry ears, women and children wearing groundhog mittens with tiny felt claws. Objects of particular devotion are groundhog amulets and groundhog key-chains. There are small whittled groundhog icons that can sit in the palm of the hand and there are 4-foot-tall groundhog tikis hewn roughly from hardwood tree trunks, carved stumps that could enchant an Easter Islander. In front of the 1936 vintage band shell stands a 12-foot groundhog ice carving. But even this pales before the "World's Largest Groundhog," a 30-foot-tall rodent totem in a top hat that rises in the parking lot of Joe's Drive-In Restaurant.

Groundhogs, like beavers, guinea pigs, and mice, are rodents. And while for many people in this country, rodent equals rat, some rodents make fine pets and in some parts of the world they make snack or dinner—the capybara of the Amazon, for instance (imagine a groundhog the size of a pig), or the muskrat of the American marshlands. During the course of my travels I've tasted both, and both, after being boiled, basted, and fried in fat, taste about the same—oily and leathery. I have no particular appetite for groundhogs, but neither do I, as do many suburban lawn mowers, devise methods electronic, ultrasonic, or semiautomatic to keep these animals from burrowing through my yard.

Besides eating, burrowing is what groundhogs do. They were doing it before humans walked the earth and will probably continue to do it when we're long gone. The Indians along the Delaware River in Pennsylvania thought the earth gave birth to the groundhog and called them *wojak,* which the colonists expanded into "woodchuck," now their proper name. These fat, reddish-brown, thick-furred, overgrown ground squirrels are a species of marmot, to be precise, and have survived in one form or another for some 11 million years. They live pretty solitary lives, are good swimmers and adequate climbers, and, when not browsing on grasses, plants, and tree bark, spend the time working on their underground network of burrows. These activities take up half their lives. They spend the other half asleep.

For an animal that lives on plants, winter in the northern hemisphere can mean sparse pickings. Sleeping for a few months is better than retiring hungry every night and a lot less work than migrating, especially if you're a lumpy creature with stubby legs. So come fall, with the days getting shorter (temperature doesn't seem to be much of a factor), the groundhog's hormones begin to make demands. The commanding hormone is melatonin.

Once the jet set's favorite (taking it induces sleep, so frequent fliers often took it to treat jet lag), melatonin was for a time touted as a treatment for everything from insomnia to depression, cancer, heart disease, and Alzheimer's. What is known about melatonin—and it was discovered only in 1958—is that it's an ancient hormone. Algae produce melatonin. So do moths, birds, amphibians, and reptiles. In these animals melatonin is secreted by a nearly as ancient gland called the pineal, which is located in the back of the head just beneath the light-penetrable skin. (This is how the pineal became known as the "third eye.") In ground-hogs, as in all mammals including humans, the pineal sits in the center of the brain.

Until the 1970s no one knew what to make of melatonin. Then researchers found that children have higher levels of melatonin than adults and that these levels begin to decline with the onset of sexual maturity. As the pineal produces less melatonin, sex hormones increase. Studies then showed that at all ages melatonin controls daily cycles of sleep and wakefulness. Produced in the evening, melatonin induces sleep. (The fact that children have more melatonin than adults may explain their ability to sleep longer.) With daylight, the pineal gland shuts down. Experiments show that the pineal is so sensitive that it will cut off the flow of melatonin after only a few seconds of exposure to light. Our daily rhythms can be readily disrupted. "If light were a drug," wrote one researcher, "I'm not sure the Food and Drug Administration would approve it."

As summer and light wane, the groundhog's pineal is on and producing melatonin longer than it's off. At a certain point, with food resources dwindling and already having fattened themselves to bursting among your plants and vegetables, the groundhog feels it's just not worth staying awake for those few hours of light each day. Besides which, the increased melatonin has depressed their sex drive (in the same way it staves off human puberty). So sometime in the late fall they wallow down their burrow and hibernate. It is no ordinary hibernation. Compared with groundhogs, bears are light sleepers. A hibernating bear's body temperature may go down a few degrees. A groundhog's body temperature drops to near freezing and its heart rate falls from 75 to 4 beats per minute. (As attuned as I am, or was, at pacing heart rhythms, one every 15 seconds is pretty death defying. I wonder if, when they awake, they suffer hibernation-head.)

A few times during the winter, groundhogs revive, have a look around, and return to their burrows until they awaken for good come spring. The question, of course, is if the groundhog lies asleep in its burrow and not in a cage in the Punxsutawney Public Library, and light can't penetrate through to the burrow, how does the lengthening day affect their pineal and its production of melatonin? The answer seems to be that the genetic switches that turn the pineal on and off have somehow "memorized" the year's cycle of light and dark. Sometime in early spring the melatonin stops flowing. When it does the groundhog's level of gonadal hormones—testosterone—increases. He or she awakens, aroused by a kind of spring adolescence that also sets birds singing, frogs calling, and perhaps our own thoughts turning lightly to love. To appreciate the goings-on in Punxsutawney, you have to think of the groundhog as a sex god promising fertility.

"The general explanation which we have been led to adopt of these and many similar ceremonies," writes Frazer, "is that they are, or were in their origin, magical rites intended to ensure the revival of nature in spring." The roots of a celebration around the reappearance of a hibernating animal go back to northern Europe. The Germans, who immigrated here and settled in Pennsylvania, brought the ritual with them. The faithful have been making the pilgrimage to Punxsutawney since 1887.

After a night of balls, banquets, kielbasa, and beer, the dedicated and the inebriated (some are both) follow giant, man-made groundhog tracks for a mile and a half up to a wooded hill known as Gobbler's Knob where, in a frigid little dell, they'll make an all-night vigil. Thinking better of spending the entire night outdoors I opt for the "Knob Bus" that begins to make its round of the nearby motels at 3 A.M., making sure its passengers arrive at the dell before dawn. By 5 A.M. we are 10,000 strong, crowded in behind the rows of chairs set up for the dignitaries. More chairs stand on the makeshift stage. A contingent of men arrives, each in top hat and black tie. Among them is the Punxsutawney Groundhog Club's Inner Circle, 15 men whose sworn duty is "to protect and promote the legend of Punxsutawney Phil." Each receives an ornate introduction mostly lost in the frigid air and to the restless crowd that rallies applause for these personages, among them this year a dentist, three funeral home directors, a drilling and pump service operator, and an insurance agent. Four of them are named Bill. The president carries an acacia cane (although there's probably not a native acacia tree within 3,000 miles of Punxsutawney) and is, by the Punxsutawney powers-that-be, vested with "the knowledge of Groundhogese." It's he who is chosen, like the Pythia at the Delphic Temple of Apollo, to interpret the oracle.

The local congressman is announced, then a state senator, then the honorable governor of Pennsylvania, Ed Rendell. Although the crowd is so dense it's difficult to see anything, word filters back that Phil's keeper has brought the groundhog in his box center stage. The Inner Circle—from where I stand a collection of top hats—pulls in close. The governor, stubby as Fiorello La Guardia, tries to stay in view of the cameras. Flashes pop. There is a commotion onstage. A cheer rolls through the crowd although no one around me seems to know what Phil has predicted. No matter, the idea is that by dint of the ceremony alone nature might be inspired to hasten brighter days and warming weather. And in the midst of a cold winter a revived sense of community might make the six weeks till spring more bearable.

I have breakfast at Joe's Drive-In, where, through the window, I can see the giant groundhog's knees, and I head back to New York City, hopeful of being on time for another pre-equinox celebration. Had I spent all night on the Knob I wouldn't have the energy for the trip. I may feel stronger, but I've found that fatigue still sneaks up on me. I've got time for a lunch break and even for my now crucial afternoon hour's nap, which I've decided to call an "Eastwood," after a character played by aging action star Clint Eastwood who finds an afternoon nap obligatory. With my daily Eastwood, and my now nightly nightcap of Jack Daniel's, I'm trying on some new rituals of my own.

This evening's celebration is one of them: I'm heading to church, for in the Christian calendar February 2 is Candlemas, one of the Church's oldest celebrations, marking the arrival of

Jesus to the Temple in Jerusalem, where Mary came to purify herself 40 days after giving birth.

As much as these two celebrations may seem a world away, they're not. In Earth's journey from winter solstice to spring equinox, February 2 comes, give or take a day or two, at the midpoint, also known as the cross-quarter. The three other cross-quarter dates come in the first week of May, August, and November. The reckoning of cross-quarters may have come west with the Celts, whose tribal origins lay in the steppelands of southern Russia and central Asia, but who began invading India and Europe in the second millennium B.C. By 100 A.D. they had spread north into Britain, Scotland, and Ireland. In the Celtic calendar, perhaps in keeping with their origins as herders, summer began in the first week of May (which is how June 21 became Midsummer's Eve); autumn and winter began the first weeks of August and November, respectively. The first week in February marked Imbolc, the beginning of spring when, on the eve of February 2, worshippers of Brigid, the robust goddess of fire and fertility, paraded with torches.

When Catholicism reached Brigid's pagans it found their rites unsupportable but too well ingrained to end them outright. Instead, the Church canonized Brigid, dubbed her Saint Brigit, and allowed February 1 to be her feast day. The convent in her name "was noted for its heathenish miracles and evidences of fertility magic. Cows never went dry; flowers and shamrocks sprang up in Brigit's footprints; eternal spring reigned in her bower."

As for the chance to ignite spring's fires? Well, the ex-pagans could douse their fires and have them, too, for Candlemas, the celebration of first light, followed the next day.

"*Fiat Lux*. Let there be light" is the first recitation this evening at the Episcopal Church of the Resurrection on East 74th Street in Manhattan. The sparse gathering of a few aging parishioners, some male couples, some mixed couples—I see no children— distribute themselves among the wood pews in the high-arched but compact cathedral. This is a ceremony for the cognoscenti. All seem familiar with the place, one another, and the service. Without direction from the royally robed priest or his white vested ministers, each knows that when the choir begins *Fiat Lux* it's time to move to the altar rail and receive a candle.

I look in the New Testament and find in Luke 2:22 that in the Temple it was the aged Simeon the Just who, like the president of the Groundhog Club's Inner Circle, has the gift to interpret the signs. The Holy Ghost speaks to him. Simeon recognizes Mary's child as "a light to enlighten." With Him, the old will pass and the new will arrive. Change is in the offing.

Candles lit, the choir and congregation sing a long Latin *Ave Maria*. The priest, who has changed from purple garments to gold, walks in procession with his young initiates. Each swing of his censer releases a voluminizing mist through which, in the candlelight, the statuary gleams. The church becomes a mystic dell.

As always, I'm amazed at all of the ritual's details. And having been educated in similar rituals—those of the synagogue, when I was young—I sense there's more nuance to them than I can know. I'm certain the priest's every movement, from the timing of his steps to the placement of his hands on the chain of the censer, came to him only after centuries of considered evolution, of priests who came before, wondering, what more can I do to influence nature, the gods, or god? Following a particularly fertile year, some priest wondered what he or his congregation did differently to bring about such favored treatment. Was it the particular mixture of spices in the censer? Was it the blue garment? Or was it the death of the old woman who we believed had the evil eye? Ritual, and much evil, evolved out of such faulty associations. For me, the fact that we now know much of the physics and nature of the world turns ritual into enthralling theatre. I enjoy its required suspension of disbelief because with the world of facts set aside and the world of the senses engaged, I allow myself to imagine an ideal of existence in which light, words, and song might alter the course of events. Out of such moments, hope springs.

The organist percusses through a Mozart piece. I hear remote Alleluias.

"The old man carried the Child: but the Child governed the old man."

Candles lit once again, the congregation sings:

> *The happy birds* Te Deum *sing,*
> *'Tis Mary's month of May;*
> *Her smile turns winter into spring,*
> *And darkness into day.*

If the groundhog sees his shadow when he arises from his burrow on February 2 (meaning there's morning sunlight), the prediction is another six weeks of winter. Just so, the folk saying goes: "If Candlemas day be fair and bright, winter will have another flight. If Candlemas day be shower and rain, winter is gone and will not come again."

February 10
10 HOURS, 37 MINUTES OF DAYLIGHT

The increasing light energizes the season and me. On my walks in Brooklyn I take the time to steep in the February sunshine that, as William Cullen Bryant wrote, "tints the buds and swells the leaves within." This is physics as well as poetics. The roiling surface of the Sun radiates waves of energy, 1.35 kilowatts of energy for every square meter of Earth. The visible energy wavelengths color the world: violet, blue, cyan, green, yellow, orange, and red. The rest, waves with frequencies a hundred million times smaller or larger, pass us unseen. All charge the atmosphere and the land.

We now build solar panels and batteries that utilize an infinitesimal fraction of this energy. But the very first solar receptors evolved some three and a half billion years ago, a time when life on Earth consisted of primordial microbes, single cells surviving on a diet of carbohydrates fueled and produced with sulfur, iron, and methane. Then came a *fiat lux* moment (a moment in geologic time) in which some of these microbes began to use sunlight to power their fuel production, at first by energizing molecules of hydrogen sulfide and carbon dioxide and then—and this is the moment when life as we know it began—using water and carbon dioxide to produce carbohydrates and release oxygen. These single cells would eventually engender virtually all the life that now exists on Earth.

When these cells, with their unlimited supply of energy, combined with other cells the results were auspicious, and in a matter of a couple of billion years plants began putting them to use. An average leaf contains 5 billion energy-transforming cells called chloroplasts. Each chloroplast holds some 600 million molecules of chlorophyll pigment. Each of these molecules, in order to capture the speeding packets of light energy called photons, has an array of upraised molecular antennae that, like narrowband radios, tune into the red and blue energy bands. (Leaves are green because it's the frequency the pigments reject and thus reflect.) They then pass the collected photons—now called excitons—from one molecule to the next and finally trap each, holding it in a molecular reaction chamber within the leaf. There, by means of a process that researchers have yet to be able to reproduce, the plant cells harness the once careening and oscillating light energy and stabilize it by forcing it to react with water. Protons are shifted, charges separated, electrons released and recombined. The result is food in the form of carbohydrates that gets stored in the plant's stems, roots, and seeds, and also oxygen that we and nearly every other living thing breathe, which gets released into the atmosphere. The whole transformation takes place in a matter of picoseconds, trillionths of a second. In fact, the reaction process is so fast and efficient that plants have had to devise ways to keep it from burning them up.

Primary among these is the vascular system, an unsurpassed hydraulic assembly that not only pumps away created proteins for storage and sucks up needed fluids and nutrients but also serves as the plant's musculature by maintaining a constant pressure within its flexible passages.

When leaves fall, when plants die, worms, fungi, insects, and bacteria feast on what remains of the stored energy and with their deaths pass it on to the soil. Come spring the budding plant takes up the soil's stored moisture, minerals, and nutrients, using them to kick-start its revival. Red light receptors in the plant's cells control budding and flowering times by keeping tabs on the relative proportion of daylight and darkness. Blue light receptors time the sun's passage and control the fluid flow.

Seeking to know more of this seasonal power surge, I pay a visit on David Mauzerall, a wiry scientist with white hair and a tight-lipped smile who has spent his life discovering the physics, chemistry, and biology of photosynthesis. From his office in the labyrinth of basement laboratories at New York's Rockefeller University, he assures me that photosynthesis is what sustains life but the reason it does, he says, is that plants, with their efficient means of production, manufacture up to 10 times more food than they themselves need to survive, even more food than they need to reproduce themselves.

"Even in the oceans," he says, "algae produce up to ten times more than is needed to support their life cycle. My question is, why do they do it?"

I must appear pretty clueless because he doesn't wait long to continue.

"The answer is, we don't know."

We know what they do and how they do it (although we can't duplicate it with anywhere near their speed and efficiency), but the why of it remains uncertain. The fact that we and other living things make use of the food that plants produce, that we even make use of the fossil fuel provided by plants that lie compacted for millions of years (as Mauzerall puts it, "To warm oneself by coal is to bask in the light of the carbonaceous"), that we depend on the kindness of vegetation, none of these is an answer. We don't know and I find myself just as enthralled by this as by what we do know. I'm open to everything.

Perhaps recognizing this, Mauzerall hands me a reprint of one of his published papers. I can make nothing of its long title. He smiles and deigns to put it simply:

"I devised a way to hear photosynthesis."

Perhaps he's spent a little too long down here in the belly of the university.

"A spinach leaf sounds different than a lettuce leaf. One species of algae will sound different from another. A healthy plant will sound different than a stressed one."

Way too long.

"I send a pulse of light through a leaf in water and the energy that's not absorbed expands in the water and creates sound waves. It's not new. Bell modulated light to create sound waves."

In 1880, four years after he perfected his telephone, Alexander Graham Bell decided that transmitting voice by light would be far more effective than electrical transmission over wires. He designed a system in which sound waves hitting a sunlit mirror were reflected as light waves. The recaptured light could then be reconstituted as sound. It worked—when the sun was out and when there was no interference. A hundred years later lasers beamed through fiber-optic cable allowed the fine control necessary to transmit sound at the speed of light.

Mauzerall's device was far simpler but its larger applications might enable scientists to tune in on the photosynthesis of a forest to find out the state of its health.

"The plant will tell you how much energy it's using," says Mauzerall.

"What do you hear?" I ask.

"I hear the whoosh of just-created oxygen." That whoosh by

whoosh changed forever the nature of the atmosphere and the life of the planet.

When photosynthesis fires up anew, the great transfer of energy that is spring begins, a conflagration of green, billions of tiny fires igniting on the surface of every leaf.

Whoosh. Do I find excitement in all of this? I do. I'm ready to travel, to chase spring.

2

Crawfish, Salamanders, and Dionysus

February 24

"Spring," a word of Germanic origin, came into English around
the year 800 and meant only a place where water comes up from
the ground, the source of a stream or river. Although the Greeks
long associated such places with the rambunctious god Pan, who,
along with Dionysus, embodied the wilder side of human nature,
the connection between the spring of the stream and the spring
of the year didn't happen, at least etymologically, for another few
centuries.

By the 14th century much more than a stream could have its
"spring," its arising, its first appearance, the place from which it
flowed or issued forth, a wellspring. Morning was the "spring of
day." Buds on a tree evoked the "spring of the leaf." Plants had a
season for their "spring, or birth." A cluster of young trees could
itself be called a "spring" and a group of birds was sometimes
called a "spring" rather than a flock.

"Spring" soon comes to mean a rising, an arising, a springing up, and a bursting forth. High tide is called the "spring of the tide." Spring speaks of buoyancy, vigor, of active power. Macbeth can call Macduff's slain father "The Spring, the Head, the fountain of your blood."

God's commandment is the "spring and beginning of all laws."

Finally, by 1547, "spring" is being used as we know it. The season becomes known as "spring." The things that grow no longer grow "in the spring of the year" but "in spring": "lusty spring," as Spenser wrote when, for Oscar Wilde, "each spring-impassioned tree flames into green." (That D. H. Lawrence troped into "The Enkindled Spring.") People could now suffer with spring fever, animals with their spring-fret. They could harvest spring wheat and spring greens and name the things that arrive in the spring— spring salmon, spring herring, and spring peepers.

Shakespeare, foreshortening the word's etymology into a single sentence, wrote, in 1591, in *The Two Gentlemen of Verona*: "Oh, how this spring of love resembleth the uncertaine glory of an April day."

Before the 16th century, though, the period of time before Easter was simply known as the Lenten season. The word "Lent" also comes from German, from the same root as "lengthen," thereby marking the time as the season of lengthening days.

I had a glimpse of the kind of isolation winter might have meant in times past when visiting the island of Skyros in the Greek Sporades, an isolated rock outcropping in the Aegean Sea where the rural life has never been changed by tourism, where people still live by what they can grow, raise, or take from the sea around them.

"What do you do here in winter?" I asked a naturalist who moved to the island from England to remove himself as far as possible from his previous life as a financial analyst.

"We celebrate a lot of naming-days," he said. "It's an excuse to get out and see other people. In fact, sometimes someone who's feeling particularly housebound makes up a naming-day just to have company to drink with."

After such solitary winters, festivals designed to revitalize community ties and share what remained of the last winter stores were welcomed. Skyros, in fact, has a spring festival unchanged in more than 2,000 years and whose Dionysian trappings still include revelers skulking the island in goatskins.

In the old Roman calendar spring began in February with the raucous festivities of Lupercalia. Named for Lupercus, or Faunus, the Roman version of Pan, the rites were designed to ensure a fertile season. At the Lupercal, the cave where the she-wolf suckled Romulus and Remus, Romans sacrificed goats—symbols of male sexuality. Priests called Luperci pressed a knife coated in the killed goat's blood to the foreheads of young boys who were then supposed to begin laughing. After much food and drink, the youths, wearing only loincloths, ran from the Lupercal and coursed through the city whipping passersby with strips of goatskin.

By the ritual's transitive properties, wives struck by the whips might be blessed with fertility. The whip, called a *februa,* became the symbol of the Lupercalia. And thus the month February.

With the advent of Christianity, such pagan festivals, celebrated throughout the Mediterranean and Europe, gained the imprimatur of the Church by dint of their concurrence with the 40 days before Easter fast, during which the observant Catholic vows to forgo some particular pleasure or another, usually the eating of meat. The Italians called this practice *carne levare*—the leaving of flesh—and so the week before the start of Lent, known as Holy Week, also became known as Carnival. Especially among the Latinate Catholics (the French, Spanish, and Italians) Holy Week became a time for indulging in drink, food, and masquerades. Lent began on Ash Wednesday; the day before, Shrove Tuesday, was officially set aside for "shriving," confessing one's sins. Since eggs and fat were forbidden during Lent, people ate them heartily on

Shrove Tuesday, which became known as Fat Tuesday. Fat Tuesday, in French, is "Mardi Gras."

Had the French settled America before the English, spring's arrival might be much more feted today. The Puritans were not a festive bunch and they were especially hostile to displays of humanity's lustier nature. The French, however, did settle Canada first. In 1604, emigrants from Poitou in west-central France landed in what is now Newfoundland and Nova Scotia and called the colonies they established Acadia, which is French for "Arcadia," which was the Greek land of peace and harmony. It was, in fact, where the god Pan was born.

After a century of battling the British over the land, the French lost and in 1755 faced exile from Acadia, which the victors renamed Nova Scotia. Thousands of Acadians moved south into the New England colonies. Some returned to France. Others took ships down to Louisiana, where the Spanish, who then held the colony, allowed the French to land. On the prairies of Louisiana north and west of New Orleans, the Acadians could farm and trap—if not seals, then beaver and muskrat. These were, and still are, pretty wild lands where the streams called *bayuk* by the native Choctaw and "bayou" by the settlers move so slowly that sometimes they're indistinguishable from the nearby cypress swamps, where, if it wasn't for the levees, land itself might be indistinguishable from water.

The prairies were a haven for outsiders, from Indians to escaped slaves and German immigrants. Among them languages and customs creolized, becoming what's known now as Cajun.

In my search for an American Lupercalia, a documentary film with the auspicious title *Dance for a Chicken* shows me that I ought to head down to Cajun country. I call the director, Pat Mire, a resident of Lafayette, Louisiana, who, as it turns out, was born and raised on the prairies. With what seems to me a very hospitable Cajun twang he says he'll be glad to serve as my guide to the Cajun Mardi Gras.

I fly into Baton Rouge and drive west to Lafayette, then north toward Opelousas. On the radio I'm already hearing the distinctive triplet-driven music that's known as swamp pop, some sung in English, some in French, and some switching readily from one language to the other. West and northwest of Opelousas, where flat waterlogged lands are divided by swamps and bayous, lies the heart of the Acadian prairie: Eunice, where I meet up with Mire, and the even smaller towns of Mamou and Basile. Spring is just arriving on warm air that in February begins to sweep up from the Gulf of Mexico. Magnolias are already in full of pink blossom and around the farmhouses bloom rubescent azaleas and camellias. Mire says they'll be planting sweet potatoes next month and rice seed as soon as it warms.

He points out the large rectangular ponds next to many of the farmhouses. These serve as both crawfish ponds and rice paddies.

"Those open boats you see in them, we call them bateau," says Mire. "We use them to harvest the crawfish. You set out in the boat and pull yourself along the trapline. The boat drifts, the farmer pulls in his rigs and traps, dumps the crawfish into the boat, and then moves on around the pond. If you're good at it, it's a kind of beautiful ballet."

This is the last crawfish harvest before spring. An important harvest since, Mire says, more crawfish gets sold and eaten during Holy Week than any other time of the year. Soon after, the water will be drained from the ponds and the wet fields planted with rice that will flourish on the nutrients left behind by the crawfish.

"We'll have two rice harvests, June and August, and then, come October, the fields are flooded up again and the crawfish come out to feed on the remains of the rice."

To the rural Acadians who take their irreligious revels as seriously as they take their holy rites, the famously raucous Mardi Gras in New Orleans is something of an abomination.

"Mardi Gras is a celebration of community," says Mire. "It's about getting together to share the wealth after a hard winter."

Each town has its own Mardi Gras run, called a *courir*, organized by the entire community, in which masked and costumed mummers on wagons or on horseback parade from farm to farm in an antic march of pantalooned buffoons. Each *courir* ends with a community *boucherie* and feast. As Mire showed in his film, the towns' traditions differ in music, style of costume, and performance. The reason, he explains, is that until bridges were built across the bayous, the towns remained pretty much isolated from one another and their traditions evolved separately. (I think of the beaks of Darwin's isolated populations of Galapagos finches.)

One thing, Mire says, if you can't make it for Fat Tuesday in Cajun country there's always Fat Monday, Fat Sunday, or Fat Thursday, and each day—Lundi Gras, Dimanche Gras, Jeudi Gras, and Mardi Gras—is celebrated with beers, *bals, courirs,* zydeco, *fais do dos, boucheries,* andouille, boudin, and crawfish boils.

Mardi Gras morning we begin with what Mire says is a traditional breakfast of beer and *boudin blanc,* the latter being a thick pale sausage of cayenne-spiced pork, pork liver, onion tops, and rice, that we pick up at Mire's favored location, a combination gas station, grocery, and butcher shop. We take a seat on the curb in the parking lot and eat, beer in one hand, slickery boudin in the other.

"It's a good start," says Mire, when we finish. Not one my heart surgeon would care to see, but when in Rome or Eunice for Lupercalia, do as the Mardi Gras does. In these towns Mardi Gras refers not only to the day and the event but also, in what seems to me a very nuanced usage, a Mardi Gras participant, since he or she who embodies the spirit of the day can also be called Mardi Gras.

The best way to observe what goes on, says Mire, is to find a place along the route of the *courir* and await the arrival of the Mardi Gras. At a farmhouse with a crawfish pond next to it we pull over alongside several other cars. About two dozen spectators stand in waist-high grasses along the roadside (most of the town is in the parade). Everyone has a beer in hand and is looking down the blacktop road that runs across the flat, open land, vanishing into a glint of sunlight.

Someone calls, "Mardi Gras!" But all I see is a glimmer along the horizon about a quarter of a mile away. Ten minutes later I can make out a line of silhouetted horsemen in high conical hats. I hear shouts and muffled music.

"Here they come," says Mire. And as if a curtain is drawn aside, they're upon us. An unmasked *capitaine* rides in the lead on an elegant Appaloosa; behind him, a drunken and gaudy dragoon of masked grotesques. The bright satin costumes of the riders aren't store-bought or makeshift but well-tailored multihued tatterdemalion—quilted, checkered, tufted, and fringed. The masks are shaped of flexible wire screen so they have no eyeholes but bear wild painted faces, some adorned with stuffed cloth noses—long, bulbous, or twisted—in blue, purple, and red. Along with the riders comes a horse-drawn wooden flatbed carrying the band that, with its accordion, fiddle, guitar, and clarinet, adds Cajun improvisations to the bedlam of color.

When the procession reaches the farm the *capitaine* holds aloft a white flag and raises a long whip of braided canvas strips (his *februa,* I think to myself), and his merry band, some hundred altogether, comes to a halt. The *capitaine* stands down and with some formality addresses the owners of the farm, a man and a woman holding a two-year-old in her arms. With the *capitaine*'s approach the child buries his face into his mother's neck.

"Le Mardi Gras demande voire permission pour visiter ta maison," announces the *capitaine* with exaggerated chivalry.

The couple must have granted the permission requested because at the wave of his flag, yelps arise from the mounted clowns, who spur their horses to a gallop. They charge across the broad front lawn, dismount on the run, and, with the music started up again, begin cavorting with the crowd.

Once again the *capitaine* raises his whip. The clowns convene in front of the homeowners and form a tight huddle, some on their knees, some standing. The *capitaine* commands them to sing and, like schoolchildren (schoolchildren who had their first beer at 7 in the morning), they sing a song in French that, Mire tells me, is a request for charity, for food, and to have their empty bottles filled. What they want most of all, says Mire, is a chicken that they'll be able to use for the evening's communal pot of gumbo.

The song ends. The farmer appears impressed. He goes to the back of his house and returns with a live chicken in hand. Cheers go up. But instead of presenting the bird to the masked beggars he throws it in the air. All hell breaks loose.

With dozens of drunken masked young men after it the chicken runs, flies, heads for the roadside brush. Its pursuers, their vision impaired by their masks, their sureness afoot impaired by beer, slip on the wet ground and fall in heaps of colored satin. The chicken is no easy prey. It dodges and ducks (if a chicken can duck), dives and evades tackles, and just as it seems surrounded, its cause all but lost, flies—flies!—up into the branches of an oak. I don't ever recall seeing a chicken in a tree and the frustrated clowns, being laughed at by the crowd, seem just as nonplussed.

In the meantime the musicians have struck up some lively zydeco and the Lite Beer van and the horse-drawn boudin wagon have arrived, the latter a flatbed holding a huge insulated chest filled with steaming sausage. In a surreal touch, three young girls in white tutus and angel halos stand on the wagon handing out the boudin. The sight of a grown man on horseback begging sausage from a fairy nymph would make a pagan Roman proud.

Those Mardi Gras who are neither dancing nor chicken chasing go through the crowd on their knees, begging miserably, all of them in the same way, forefinger scratching at their upturned palm and pleading, *"Tit cinq sou? Tit cinq sou?"*

A few young men in brave displays of beery horsemanship stand on their saddles, beer in one hand, reins in the other. Two revelers grab a child and begin a mock kidnapping with the *capitaine,* whip in hand, in hot pursuit. When they're caught they submit to their canvas lashings with abject humiliation.

Meanwhile, the chicken chasers have found a ladder, leaned it against the tree, and are climbing unsteadily up toward the chicken. It's a Keystone Komedy waiting to unfold and it does. The chicken flies. The ladder falls. The clowns spill across the lawn. Finally, the chicken is caught, raised high by its feet, and put into a cage with the other birds caught that day.

The rout seems ended but one of the clowns lies motionless on the ground at the foot of the chicken tree. Another masked mime kneels by his side. He waves the others over. Soon the lot of them line up like graveside mourners and keen. The *capitaine,* wielding his whip, comes over and presses through the group. Looking down at the very still masquerader he calls for a beer. It is quickly brought. He kneels, pulls open the clown's mouth, pulls the tab on the beer can, and pours. There's some gurgling and coughing and then to the cheers of the crowd the clown opens his eyes, smiles, and arises. Resurrected! The Mardi Gras brings a man back to life! It is, says Mire, a traditional passion play that anticipates the revival of the season and the resurrection to be celebrated 40 days hence.

With a call from the *capitaine* the Mardi Gras mounts up, some riders struggling up into their saddles, and the parade, the band, the boudin nymphs, and the beer truck head off down the road.

It's only 10 in the morning. I ask Mire what they'll do the rest of the day.

"They've got many more farms to go," he says.

"I assume they'll sleep late tomorrow."

"Not tomorrow. Tomorrow morning they'll all be in church."

What better way to prepare for the holy weeks to come than by mocking all that is held sacred and holy? A "wild association," as Cicero described Lupercalia in Rome, "both plainly pastoral and savage, whose rustic alliance was formed before civilization and laws . . ." I find this ritual refreshing in a country that more and more takes its Gospel too seriously and shuns what is pastoral, sensual, and savage in the soul.

But the Cajuns themselves call their prairie world a *petit monde,* a separate place where, for many years, they faced cultural discrimination, stigmatized as rural curiosities. Mire recalls how parents were discouraged from teaching their children French. Those who could speak French spoke it only among themselves, never in front of strangers or in public. As is often the case, the isolation only strengthened their cultural identity and resolve. It seems to me that no one could celebrate spring better or more authentically.

March 7
11 HOURS, 33 MINUTES OF DAYLIGHT

When I return from New Orleans I am blessed with a car by way of my friend Miles, a Heidegger scholar and professional wrestling fan. Miles's mom died a few months ago—at nearly 90—and left him her car, a car he says she'd kept garaged and used only for shopping and church on Sundays.

I think I know a Heidegger scholar from a used car salesman, but I've got thousands of miles ahead of me and so I tell him it sounds great but will the car make it across the country, through the desert, across the Plains, up the Rockies, and back? What kind of car is it?

"A Chevrolet Impala."

"What year?"

"1984."

"Let's go have a look."

But the car is garaged at a friend's house in Connecticut and Miles won't be able to get the key until the end of the month. More waiting.

For the last month I've been alternating working out at the gym and taking four-mile walks around Brooklyn's Prospect Park. I'm feeling pretty good and when I have my pretrip checkup, the cardiologist says he's ready to make me poster boy for heart valve surgery. The valve's working fine. All vital signs are vital. I tell him I plan to go across the country, hike in the desert and in the

Rockies, and fly in a small plane to the Arctic to hike some more. I look to see if he shows any sign of unease but he says he has none.

"Altitude no problem?"

"If you feel tired just stop. You should be fine."

I'm thinking he might show at least some concern but he just smiles and says, "Sounds like a great trip. Call me when you get back."

After months of feeling tethered to my recuperation, I have no more excuses. I'm good to go. I'd been saying, "Come spring . . . ," for so long—now it's here.

March 21

12 HOURS, 12 MINUTES OF DAYLIGHT

Anxious to get under way but waiting for the car, I decide this first day of spring to meet up with my friend Michael Klemens, a biologist with the Wildlife Conservation Society, who has fieldwork to do in a forest north of New York City. I take the train and meet him at his office in Rye, where we pack his Jeep with field gear and head north. For a herpetologist such as Klemens, the winter, when reptiles and amphibians lie holed up in their hibernacula, can be a trial. I've heard him grouse like a ship captain confined to port about the unrelenting dormancy of the season that leaves him too much time to contemplate dark thoughts, to the point that he sometimes even rues the coming spring for the fact that when it finally comes he'll see how much more of his woodlands and wetlands, his animals' few remaining safe havens, has been lost.

So Klemens looks for signs of spring sooner than most.

"There comes a time around the end of February when I really start to get a quickening because I see the days are markedly brighter. I start to see the buds swelling, or I think they're beginning to swell—wishful swelling—and I start to think, Well, I've got to get my minnow traps and dip nets ready. And I watch the jet stream. I watch the weather patterns. I start to prognosticate. I begin to think salamander thoughts."

Several years ago, Klemens began seeing the rural lands that held the last best hope of survival for the species that he studied being overrun by suburban sprawl.

"They were moving, relentlessly, into my refugia," he says. "Humans seeking new habitat, following the same paths as the amphibians did after glacial retreat, moving up the river valleys, claiming new breeding grounds, bringing roads and more of the same yuppie genotypes with them."

In this period the population of the New York–Connecticut corridor was increasing 10 percent while the amount of land taken up by development was increasing by 60 percent, each new sprawling subdivision spawning new roads, sewers, schools, and shopping malls, the stuff of salamander nightmares. Klemens decided that one way to address the problem was to educate local political leaders in ways to protect the environmental integrity of their small towns when they came up against big developers.

"Development's going to continue," he says. "Humans are on this planet for better or for worse. It's a matter of *where* we develop and *the types* of development. Traditionally, there're two ways of dealing with land. Lock it up and protect it in preserves or say we're developing this area and basically giving it up. The idea that preserves alone can protect nature is a myth. We can't put away enough land to really protect ecosystems' function and integrity. We can't lock up enough land. We don't have the money or the resources. I realized I needed to change the conversation among conservationists and developers."

On today's field trip he plans to survey the populations of forest frogs and salamanders in Dover, New York, in order to determine the area where building will do the least habitat damage.

"What I'm saying is that the huge portion of land that we're giving up to development—much larger than anything we're protecting—has the potential to still function for wildlife."

He's a very intense field companion, more aware than anyone I've ever been out with to sound as well as sight. I've seen him suddenly dive into high grasses and come up holding a large snake that he says he heard as we were walking. No matter when you meet Michael you always catch him in midsentence since he seems to have several conversations going on in his head at the same time. The best thing to do is just find a wavelength and tune in.

For instance, when I tell him I'm interested in the transfer of light energy from sun to plants and animals in spring, he doesn't hesitate long enough for me to get out my notebook.

"Spring? Energy? Think about all the leaves in this forest," he says, despite the fact that we're not in the forest yet. "They're all energy. All these deciduous leaves. They fall on the forest floor and they sit there for several years and eventually decompose, broken up by wetting and drying. All the organic material, all the sun's energy, is released into nutrients.

"In comes the wood frog, in comes the salamander, the young suck up all the nutrients and transform. By the thousands and thousands and thousands—they take all that energy that came out of the leaves and transfer it back into the ecosystem. The leaf fall feeds the frogs. Everything eats the frogs. So where you see little froggies hopping around, I see energy hopping around."

On spring's first day, the forest's altitude is high enough that cold air still hangs in, damp, vaporous, virginal. Dead leaves lie a foot deep. Our boots break through some remaining patches of sun-glossed snow to softening mud beneath where young vines already sprout bloodred tendrils. Near the summit of the slope we come upon a shallow swale filled with dark water—a vernal pool.

"Vernal," from the Latin *ver,* means "of the spring." A vernal pool is a depression in the forest that fills in spring with snowmelt and rain. If you're going to think salamander thoughts, this is the place to do it. In the dark of early spring nights when hard cold rains fall, amphibians—frogs and salamanders who must lay their eggs in water—arise from their winter burrows and, as full of bearing and ardor as any dam-climbing salmon, rush by the thousands through the barely thawed woodlands, even beneath the snow, even across rainswept roads, to reach these, their ancestral pools.

Females meet males in the ponds and salamander courtship ensues. The pairings begin with a male's tentative embrace (known scientifically and, I think, more sexily, as "amplexus") with the male, say, a shimmering black Jeffersonium salamander, pulling his way up the female's back until he can clutch her behind her forelimbs. Thus embraced they may drift for minutes or more but soon the male begins to rub, first his snout along his mate's head and then with flicks of his tail, his body, back and forth, and presses his cloaca against her. In the throes of his ardor, with his tail lashing, he releases his grip and drifts up and apart from his mate. While he thrashes in his frenzy she approaches and presses her snout between his lower limbs, which excites him into releasing packets of sperm. With swift movements she picks them up as they fall and places them within her own cloaca in order to fertilize the 200 or more eggs she holds inside. A day or two later she'll release the eggs, small as shot pellets and bound up in protective translucent gelatinous clusters that cling to sticks and grasses in the pond.

At this same time the eggs of tiny fairy shrimp that, dehydrated and frozen, survived winter in the dried-up swale begin to hatch with the snowmelt and rain. The shrimp eggs will serve as food for the young salamanders, for by the time they're loose from their own egg mass the parents will have retreated back into the forest.

By late spring the vernal pool, evanescent as a mirage, has dried up.

Michael stalks the pond's perimeter for signs of amphibian coitus but by his figuring we've missed the show. He sets out wire minnow traps that he'll pick up later to see how successful the spawning has been and we hike downhill to the next pool on his map.

We hear this one before we see it. The stammered cluckings of wood frogs rise up the slope from dense thickets below, standing half-submerged by spring rains. At first I see nothing in the dark pool, then Michael points out, perched on the bank, a palm-size dark gray ridge-backed frog with black mask stripes across its eyes. There must be thousands of them here, he says.

And then I begin to see them, rushing away from our footsteps, sinking into the water so that all we see are the oily gleams of hundreds of pairs of black, beadlike eyes.

The winter hibernation of these frogs has been far more death defying than even that of the groundhog. Only a week or so ago these frogs sat beneath the leaf litter in cryogenic suspension, two-thirds of the fluid in their bodies turned to ice—no blood flow, heartbeat, or breath, no brain function. The only thing between them and death being a continued slow metabolism producing enough proteins and glucose to keep their cells from drying up and freezing. This ice age adaptation allows wood frogs to live farther north than any other frog—as far north as the Arctic.

They literally thaw out in spring, heart first, in only several hours, and begin life anew. These frogs, I think, may be my *mannitto,* the thing in nature that the Delaware Indians believed reflected a person's spirit.

"Listen," Michael says.

A shriller quack pierces the miry air.

"A panicked female," he says.

I don't wonder. It's been a long cold lonely winter.

The amphibians resolved with some elegance the problem of moving from birth in the water to an adulthood lived on land by dividing their lives between the two. But they can't live without both. In a few minutes a bulldozer can bury one of these little vernal pools and render moot 300 million years of evolution.

"If you preserve the pool but lose the leaves that fall into the pool," says Michael, "the pool is useless. You've stripped them of their energy source. And if you lose the upland habitat the animals need when they're not in the pool, the pool is useless. It's terrible. We underestimate the importance of what together all these things mean."

How can we know what they mean unless we're aware of their existence? All of these things together: light, leaves, soil, pond, water equals salamander. How many of us will see any of them this spring or any spring? The fact is that most of spring's grandeur—each day's increase of energy—goes by as unnoticed as the sounds of photosynthesis.

On our ride back, Michael lectures me on land use.

"The problem today is that we have a very narrow view of land. We view land as a construct we hold at the moment without concerns for where it comes from and where it's going. We've basically discarded the public interest, the commons, the obligations of stewardship and just have the 'me' obligation—it's my land to do with it as I please. . . ."

But I'm not listening closely. This first trip of the spring has tired me out. And anyway, my mind is still back in the forest, where spring began.

March 31
12 HOURS, 40 MINUTES OF DAYLIGHT

Miles and I head up to Connecticut to retrieve the Chevy. It is at a small summer cottage with a large garage. Miles has a rusted key for the rusted padlock. We open the lock, swing out the large garage door, and even in the dull shadows I can see that this is not just a car. A Heidegger scholar doesn't lie. This is a classic. It fairly takes my breath away. This is huge. Really huge. A white body four-door with a 1980s slant, the front end riding higher than the rear. I move aside a couple of bicycles and a folding table and open the car's big door. The roof light comes on and I peer inside. It's got a cavernous interior of cobalt blue cloth, a hood that stretches out to its own horizon, and a trunk that I'm sure can hold both being and nothingness. Nothing digital here; nothing automatic except the transmission. The radio has push buttons and you have to muscle down your own window if you want to adjust the side-view mirror. The car has a bench seat big enough to sleep on, a dash shelf a foot deep, and a glove compartment that could house a large cat. But most amazing of all is the odometer. In 20 years the car has traveled only 28,000 miles. I lift the hood. The engine is a museum exhibit. We used to say when we turned on the ignition that the car "came to life." We said this because cars once looked and felt like this one—big machines into which we mortals liked to feel we breathed life. This car doesn't have a trace of ergonomic sissification. No cruise control, no lumbar support, and no cup holders. This car is yearning for a new life, to be out on the open road. Miles and I decide to call him Dick, after Moby. I take the key, turn it, and the big white whale of a car comes to life.

I back him out of the garage and in the light he seems even bigger. I can't wait to get out on the road. On my way back to the city he drives fine, picks up slowly but once going has the inertia of a runaway locomotive. When I take my foot off the gas, there's not a hint that he'll slow down. On my first quick turn the back end fishtails and I realize that until I'm used to his size, I'm going to have to mind my, or his, physics.

April 2
12 HOURS, 45 MINUTES OF DAYLIGHT

I'd say rain is threatening but since one of my few rules for the next three months is to deal with whatever weather comes, I begin loading up the car. My other rules are the following:

ALWAYS HAVE ENOUGH GAS
LET YOURSELF BE DISTRACTED
YOU'RE NEVER LOST, JUST EXPLORING
FOR HEALTH, TAKE ONE EASTWOOD A DAY
AND ONE JACK DANIEL'S EACH NIGHT

The dimensions of the Chevy's trunk simplifies the packing. I pile in a tent, sleeping bag, camp gear, carton of books, carton of clothes, coats for cold, rain, and snow, boots for hiking, a pair of waders, cookware, and a cooler chest. No arranging required, no trouble closing the trunk. Maps and guidebooks go in the glove compartment; notebooks, binoculars, compass, box of pens, tape recorder, and cassettes go on the seat beside me.

I leave in a cold downpour and head west for Allentown, Pennsylvania, where, in January 1950, I was born and where I lived until I graduated from William Allen High School. It had been a small but growing city, an industrial town home to Mack Truck, Neuweiler Beer, Freihofer's Bakery, and Hess's Department Store, as well as my grandfather's Bru-Mar Manufacturing Company (named after my brother, Mark, and me) along the Little Lehigh River, where he produced women's and girls' bathing suits. For him, as for many of the Jews in the local garment businesses who immigrated to the United States and worked their way out of the dismal Lower East Side of Manhattan, Allentown was the Promised Land. A train still ran a regular route to New York City, the track running east through Bethlehem and its steelworks, and Easton with its Delaware River industries, and through broad rolling farm- and pasturelands.

In 1950 the city was just on the cusp of its suburbanization, and now I want to see again the little two-story house at the end of the Vine Street cul-de-sac where we lived for my first eight years, before we moved out to the L-shaped ranch home in the nascent suburbs.

That cul-de-sac was full of young families, and my grandparents lived just a few doors away. Since few cars came in or out we would circle around on our bikes or play ball as late as the light would allow and sometimes longer, until in the dusk the ball would become a soft apparition of a ball, and even until you couldn't see the ball at all and just put up your glove in the hope it would appear.

Come spring the Little League held baseball tryouts on the still-muddy field, a field with real outfield fences and limed base paths. Behind the high school we'd play a rough game of two-hand touch until my friend Dutch's mom sent his German shepherd out to bring him home for dinner. The dog didn't wait for the next play. Every pair of Dutch's pants had ripped cuffs.

Weekend mornings, instead of waking my parents, I'd traipse over to my grandparents' house—still in my pajamas—knowing that my grandmother would make me breakfast. She would tear a hole in the center of a piece of bread, butter the frying pan, put the bread on the pan then crack an egg so the yolk dropped right into the hole. The idea of combining toast and egg in this way seemed to me at the time to smack of genius.

In the backyard of our house was a willow tree, and though I couldn't climb high enough into the tree to see into the window of my second-story bedroom, from my bedroom I could look out the window into the tree's weeping boughs. I recall how bare that tree looked in winter, sometimes trimmed back near to its trunk. And then one morning I'd look out and it was full again and bright green. I guess that because I went to school in the morning and in the winter went to bed when it was already dark, I missed the tree's leafing out, which I realize now must have happened quickly since it stood on the south side of the house. This may also be why I remember my room as always bright. But then, these were bright years.

I just want to see again the place where, as Old English would have it, I spent the spring of my life.

I find the little cul-de-sac pretty quickly. The street is clean. The homes look a bit run down, but otherwise the little neighborhood of two-story brick twin homes appears no different. I park the car, walk over to the house, and, with rain falling, place myself on the sidewalk so I face the front door. The house appears as in a photograph, its dimensions foreshortened by the haziness of both the day and my memory. The modest front yard seems even smaller. The sidewalk needs repair. The coal chute window to the basement is boarded up. I remember the truck pulling up to the house, the coal man's soot-smeared face and hands as he nudged the chute from the truck to the window, and the bright black cascade of coal as it rattled down to the basement.

No one seems to be home. Anyway, the house doesn't look very welcoming, or it may be that I'd rather not walk into my photograph. Perhaps on a sunnier day. Instead I walk a couple of steps to have a look into the backyard and for a moment actually think I see the willow tree, but it's not there. I don't know what I expected. It would have to have been a very long-lived tree.

I get back into the car, take a last look before I drive off, and head out to find, happily, that the old Philly steak sandwich shop where my uncle used to take me when he baby-sat is still here. It's still called Zandy's and still run by the family of Ed Zandarski who opened the place in 1940. The menu, which still fits on a plastic signboard, hasn't changed: steak sandwich, cheesesteak, hot dog, Taylor pork roll (I still don't know what it is), hamburger, French fries, pirogies. I order a cheesesteak with everything, Coke, and fries and read the old clippings on the wall. The willow tree's gone but Zandy's cheesesteaks still come in a sheet of wax paper and nested, as ever, in a red plastic-weave basket. The taste of griddle-fried flaked steak, fried onions, melted cheese, tomato sauce, and pickled hot peppers—for a kid from eastern Pennsylvania it's as good as tea and madeleines.

April 4
12 HOURS, 50 MINUTES OF DAYLIGHT

Spring's timing depends on the particulars of place. While climate may rule the overall scheme of temperature and precipitation, weather, like politics, is local. When I left New York, the magnolia tree in the south-facing front yard had already burst into fragrant fulgent pink and white while the neighbor's backyard tree, facing north, still bore only buds. Where snow comes early and remains, it can insulate the ground beneath from freezing solid. Come spring such ground may thaw sooner than the ground in those places where no snow fell at all. On this drive west across the Susquehanna Valley and then south through the rolling low hills of western Maryland, I've been seeing spring in various phases, from fields freshly plowed to pastures already greening.

Although my first destination is North Carolina, where I'll be meeting with ecologists at Duke University studying the effects of carbon dioxide on forest growth, I decide to follow the western ridge of the Appalachians before cutting back east. I'm in no rush, but I find myself driving hard, feeling as if I need to put distance between home and myself. I take tentative detours onto side roads, stop to watch a hawk circle or some deer graze, drive through a couple of small towns, but spring hasn't yet arrived and I don't yet feel that I'm on its track. I try to think up an observational methodology—perhaps looking for certain species or marking forest changes—but I decide that methodology is just what I don't want. I have a final destination, the Arctic, and have three months to get there. Everything I need is in the trunk of the car. I'm going to wait to see what I find and then what I feel. I promised myself this time and I'm going to make the most of it. I'll recognize spring when I see it.

And anyway, when it comes to spring, events mean more than calendar dates. To the ancient Celts spring began on February 2. And one could make a case that it's at that cross-quarter that the increasing radiation of the Sun actually initiates the change of seasons. Plants and animals evolved molecular memories to track the rhythms of light and time. So long as they remain within their climatic ranges, such "memory" serves them well. What makes humans so adaptable is that we're not bound to a single climate by our molecular clocks and internal rhythms. We can learn the nuances of a new climate, share the knowledge that proves valuable to our survival, and pass it on from one generation to the next.

Prehistoric people developed very detailed understandings of seasonal change. Those who lived nomadic lives herding, gathering, or hunting needed to know when to move, how long to stay, when the fruit would be ripe, where the animals had gone, where they were going next. But since weather could be so variable from one year and one place to the next they needed to know what particular changes in one place could tell them what was happening somewhere else. For instance, based upon tens of thousands of years of such knowledge, the aboriginal people of Australia divided a year into anywhere from 5 to 10 seasons of four weeks' to a few months' duration. The divisions were fluid, that is, they were based upon an observed change in winds, weather, or the behavior of animals.

People of the Arctic, whose histories go back 8,000 to 10,000 years, "perfected their oral knowledge of nature's calendar," as one researcher put it, an ability to make connections between one change and another. A volume of climate studies edited by Mark D. Schwartz, a geographer at the University of Wisconsin, contains more examples: The Blackfoot of Alberta went out to hunt bison bulls when the golden bean flowered. When salmonberries ripened, the Nuu-Chah-Nulth of Canada's west coast knew the sockeye salmon had begun their run up into freshwater streams. The Stl'atl'imx of British Columbia went to collect cedar roots and basket grass when they saw wild roses in bloom. When ocean spray bloomed, the Comox, also from the Pacific Northwest, went digging for butter clams.

For many people these signs and portents were how the gods communicated with humankind. And it was clear by a man's successes at hunting, fishing, or farming that the gods favored those who were most attuned to their signs. "Learn a lesson from the fig tree," says the Gospel. "Once the sap of its branches runs high and it begins to sprout leaves, you know that summer is near." It could also be, and it's my hope for this journey, that recognizing the changes in the natural world, one might recognize something about one's own life. To the ancient Chinese poets and monks who traveled across their country, nature was nothing but signs and portents. The poet's challenge, like that of the nomad's, was to recognize them.

Whether the gods became unreliable or those who understood the gods' signs disappeared is not known, but at some point people began keeping written records of weather and of planting and harvest times. The first such almanac may be a 3,000-year-old Chinese calendar called Xia Xiao Zheng. In Japan, in A.D. 705, the royal court of Kyoto began recording the date the cherry trees blossomed. Aztec codices from the 13th to the 15th centuries chronicled in colorful pictographs periods of drought, rain, and cold and their effects on crops and human fortunes. In the Middle Ages, European estates kept agricultural chronicles. But the scientific study of timing of natural events began during the Enlightenment, in England, to be exact, in 1736 when Robert Marsham, a sharp-nosed, powder-haired landowner from Norfolk began perusing his estate with the purpose of recording "Indications of Spring." With the first signs of winter thaw, Marsham ventured out to note when his oaks leafed, when snowbells and butterflies appeared, when frogs first croaked and mated, when migrating swallows returned, when cuckoos called and rooks nested. He persevered in this rigorously annotated annual chronicle until his death in 1797 at the age of 90, when his family took over the record keeping, which they continued until 1947.

For his efforts, Marsham, the precocious amateur, was dubbed progenitor of the science of phenology, the study of the timing of recurring biological events. Based on direct observations—the Greek word *phaino* means "to appear"—it's a science of things seen and recorded. By systematizing the evidential record, a phenologist (not to be confused with "phrenologist") hopes to recognize trends and relationships between, say, the time a tree's leaves appear and the time the moth that feeds on the leaves hatches. For the phenologist, timing is everything. Of course, to gardeners, farmers, and Australian nomads, this wasn't news, but the development of long-term data for many plants and animals over an expanding network began to bring sense to everything from forest structure to bird migrations.

Over the next century and a half, many European scientists found the record keeping worthwhile, fitting for the changing view of the world brought about by James Hutton's Uniformitarianism, Charles Lyell's *Principles of Geology,* and later, Charles Darwin's *Origin of Species.* These works, among others, made it increasingly accepted and acceptable knowledge (that which had long been common knowledge in Asia) that the natural world was and always will be in the midst of transition; that change, not stasis, is the norm. Phenology even had a kind of faddish following. After all, one could chronicle the natural history of one's estate only if one had an estate. And what better way to make an afternoon walk appear useful. Needless to say, not all the diarists were as rigorous as Marsham and there was much fair-weather phenology.

The pursuit seemed a bit effete to American naturalists. Watching for a tree's first leaves or a flower's first buds didn't hold the fascination of exploring unmapped wilderness. Even when American scientists, led by agricultural practitioners, recognized the importance of phenological records, the recording of them tended to last only so long as the enthusiasm (and funding) of the researchers. In 1851, the Smithsonian Institution began tracking 86 species in 33 states but the effort lasted only until 1859. Even the U.S. Weather Bureau failed to maintain continuous records. Finally, convinced that phenological observations of weather patterns could improve predictions of crop yields, in 1957 J. M. Caprio at Montana State University deputized 2,500 volunteer observers in 12 western states to keep their eyes on the budding and flowering times of purple lilac plants. It was a start.

In the 1970s, with rising CO_2 levels leading to predictions of greenhouse effect and global warming (which the media and many scientists have since been cowed into referring to as "climate change," which is like calling an earthquake a seismic adjustment), theoreticians said that on their computer monitors they saw frightening visions of the future. While most scientists agreed that global warming was real—from 1900 to 2000 the average global temperature increased 0.6 degree Celsius—skeptics could point out that no one had ever seen evidence of its effects.

But the evidence was there in the phenological record. You could look it up. In England, where they've continued Marsham's gentlemanly pursuit, much of his "Indications of Spring" had already changed. Oaks in Surrey now leaf out an average of 20 days earlier than they did in 1950. In the 1990s in the West Midlands, observers began spotting the orange-tip butterfly 12 days sooner than they did in the 1980s. Analysis of the data suggest that the peak of the British butterfly migration will occur anywhere from 2 to 10 days earlier with each degree Celsius of warming. A study of 20 breeding birds in Britain found they're laying eggs earlier than they did 25 years ago.

In North America tree swallows now lay eggs 9 days earlier than they did in 1959. In western Canada spring bloom comes a month earlier than it did 100 years ago. In Europe the growing season begins 8 days earlier than it did in 1970.

Are these isolated deviations? Worldwide studies of more than 1,500 species show that frogs mate, birds nest, and trees bud on the average more than a week earlier than they did 50 years ago. Making use of Caprio's lilac data, University of Wisconsin's Schwartz found that between 1959 and 1993, the first leaf date had advanced 5.4 days and first bloom appeared 4.2 days sooner.

Chief among several causes of this warming is an increase in atmospheric carbon dioxide. When John Marsham was walking his estate the carbon dioxide level in the atmosphere was some 270 parts per million. It took 200 years for the level to increase to 300. It has taken only 50 years for the level to increase to 380. These are the average numbers. In the spring, when plants use CO_2 for photosynthesis, the concentration falls. Then it rises again, reaching its high point in February, for the same reason the level falls in the day and rises at night.

But no matter what season you measure it, CO_2 levels have not been this high for at least 20 million years.

April 9
12 HOURS, 57 MINUTES OF DAYLIGHT

Rain has been hung up over the ridge of the Appalachians and the last few days I've been driving through mountain mist. When I check my chart of sunrise and sunset times I realize that in going south I'm *losing* minutes of daylight, Earth's tilt providing more daylight farther north. I decide to keep track of local hours. If I were in Pennsylvania now instead of West Virginia the sun would have risen at 6:50 and set at 7:52, giving me 13 hours and 2 minutes of daylight. Instead sunrise here was at 6:59 and sunset at 7:56.

Finally, though, a gusting cold front breaks up the clouds and for the first time since I began this trip the sky clears. Moby Dick's windshield wipers, brittle from storage, need replacing, so I pull into an auto parts shop where a thin young guy with short blond hair and large hands meets me at the counter.

"Yes, sir."

"Wiper blades, please."

"What kind of car?"

"Chevy Impala."

"What year?"

"'Eighty-four."

"Really?"

"Yeah. Big white whale out there."

"Nice. How many miles you have on it?"

"Twenty-eight thousand."

"You serious?"

"That's it."

"Mind if I have a look?"

"No problem. C'mon out."

His gray pants are too wide and too short. He comes out to the car with me and stands at a bit of a distance, ogling, and lets out a soft whistle.

"This is great."

I open the door.

"Here," I offer. "Have a look inside."

He bends his head down, looks at the dash then the backseat, and comes out shaking his head.

"Great car."

"Thanks."

"You ever want to sell it . . ."

"Not yet. I'm heading cross-country with it."

He stares longer. I imagine him imagining himself in the driver's seat. Great date car—no bucket seats—she can belt herself in right next to him. The right car can make a young man's fancy turn lightly to thoughts of love. I hate to break his revery, or mine.

"Can you replace the wipers?"

"Oh, sure, I'll do it."

I'm not above being proud of my car. M. Dick is not your effete foreign job nor your family van or your SUV. He's a car's car, a journeyman's car. No one has to know he belonged to a little old lady who never drove him over 50 miles per hour. He's getting a new start. I feel between us the spring of a beautiful friendship.

Not much spring otherwise on this cold Good Friday. At midday the temperature's barely above freezing. I've crossed the Eastern Continental Divide, which means that the rivers no longer run east to the Atlantic but run west toward the Ohio and the Mississippi. These West Virginia rivers run tortuous courses, meandering through a rumpled, counterintuitive landscape. Small mountains arise where it seems to me there should be valleys. The valleys seem to descend below the bases of the mountains. These are old rivers and old mountains, some of the most ancient land on Earth—crumpled up by the crush of continental plates hundreds of millions of years ago and eroded into their present folds—but a landscape never furrowed and deformed by glaciers. Because of this the land's rich with species that sought refuge during the ice ages, species that have since survived in cool mountaintop refugia.

West Virginia is a rough country of small farms whose pastures hang like hammocks between steep hillsides. In the 1890s when coal was king, here were coal barons with money and immigrant miners in crude camps. I pass through Clarksburg, birthplace of the gnarly Confederate general Thomas J. "Stonewall" Jackson, but also where each September the descendants of immigrant miners and railworkers hold an annual Italian-American festival.

Home rule country: Each home seems to have a candidate's poster on its lawn—Matthew Bragg, Susan Legg, Jim Paxton, Jeff and Robert Boggs. A man by the name of Buckshot Butcher is running for sheriff.

I drive up and down humpbacked hills, through moist ravines and sunken meadows. Along the river, trees are just beginning to green. Water seeps down the faces of hillside rocks. Tulip magnolias are in full snowy blossom. Clusters of lilacs bloom along the road.

Church country: On the hilltop above each hamlet a white steeple gleams. The rent-a-sign outside the church reads, "Need to do spring cleaning? Let God start with your heart." I can't find a radio station not preaching to its unseen choir. The Reverend David Wilshire addresses his radio audience on the evils of Jesus rock. Two pickup trucks pass by, carrying church pews. But coming in to Weston I'm so dumbstruck that I almost drive off the road.

High on a hillside stand three tall white wooden crosses. On each cross, arms outstretched and held by straps, bare feet on small pedestals, stand young shirtless boys clad only in white loincloths. At the base of each cross stands a young man costumed as a Roman legionnaire. And kneeling before them all is a young woman in white.

There's no place to pull over so I make a quick turn into the shopping mall that stands on the rise just opposite the mock Calvary. It so happens that the best view is through the large glass wall of a McDonald's, where I find most of the customers paying no attention at all to the crucifixion going on just across the way. They sit with their kids at tables full of cups and wrappers and eat. Even those watching the pageant don't seem to share my astonishment and I'm feeling wary about being seen to stare too intently.

"They must be cold," I hear, and there's general agreement on that.

The hill stands above a Masonic cemetery, where visitors stroll among monuments claimed with fresh flowers. At the top of the hill, 50 yards behind the crosses, stands a white unleased billboard that reads, "AVAILABLE." It is, I think, a Fellini-esque touch. But truly, I don't know what to think.

I mean, this is *the* spring ritual. Jesus, Dionysus, Adonis, or Attis, all suffered awful deaths and were resurrected come spring. All were born by miraculous virgin conceptions and represented both divine Father and divine Son. Attis's death was celebrated with frenzied sorrow and ritual bloodletting. A burial was then held for the god.

But when night had fallen, the sorrow of the worshippers was turned to joy. For suddenly a light shone in the darkness: the tomb was opened: the god had risen from the dead; and as the priest touched the lips of the weeping mourners with balm, he softly whispered in their ears the glad tidings of salvation. The resurrection of the god was hailed by his disciples as a promise that they too would issue triumphant from the corruption of the grave. On the morrow, the twenty-fifth day of March, which was reckoned the vernal equinox, the divine resurrection was celebrated with a wild outburst of glee.

Frazer (choosing his words and syntax carefully) suggests that "there can be no intrinsic improbability in the conjecture that motives of the same sort [those that put Christ's birth at the same time as the birth of the pagan sun gods] may have led the ecclesiastical authorities to assimilate the Easter festival of the death and resurrection of their Lord to the festival of the death and resurrection of another Asiatic god which fell at the same season." That is to say that "the Church may have consciously adapted the new festival to its heathen predecessor for the sake of winning souls to Christ":

> If the vernal equinox, the season at which in the temperate regions the whole face of nature testifies to a fresh outburst of vital energy, had been viewed from of old as the time when the world was annually created afresh in the resurrection of a god, nothing could be more natural than to place the resurrection of the new deity at the same cardinal point of the year.

Dionysus suffered as greatly as any of these gods, and his worshipers reenacted every moment of his suffering. Because Dionysus was represented as a bull, the devout tore a living bull limb from limb and then drank its blood as if they were drinking the blood of the god himself. And then they reenacted the resurrection of the god that brought spring.

The insouciance of onlookers and passersby gets to me. A few drivers pull over, get out, take photos, and drive off. Otherwise, the kids aren't drawing a crowd. Yet for me—what is it? It is that for a Jew the image of the Crucifixion evokes centuries of horrors committed by those with that image in their minds. I can't look at it with equanimity. At least the death and life of Dionysus led to feasting, drinking, and a parade of giant phalluses. Spring was the time when Greeks and Romans opened the jars of wine set to ferment in the fall, to "drink of the lifeblood pressed from its flesh." Even the dead were invited to drink with the living for they had "descended into the Earth from which new life is born."

I have to admit the kids have done a great job with their performance and I imagine, or hope, they had fun getting together their costumes and developing the scenario. Performance is gratifying. It was to the medieval guildsmen who each spring prepared their brief biblical pageants. (Think of Bottom and the mechanicals in *A Midsummer Night's Dream* playing out Pyramus and Thisbe.) It is to the Mardi Gras participants. Dionysus was not only the god of spring; he was the god of theatre, of performers, of the re-creation of life.

Performance is its own kind of ritual. Ritual is its own kind of performance. I've known both. By the time I was 12 I'd read the Bible in Hebrew. I'd studied some Talmud, learned every prayer and every ceremonial nuance so that by 13 I could, and did, lead daily and Sabbath prayers for the congregation, reenacting the rituals I'd been taught, performing them as I'd seen them performed. I questioned many things about my life, but in feeling knowledgeable in these rites I felt a kind of competency difficult for a young person to come by. I felt a sense of community, not religious fervor. I felt connected to my grandfather more than to God. I loved to chant the prayers with him and sometimes even now, when I'm driving or walking, some lovely ancient melody will come into my head and I'll sing it in Hebrew to myself.

I might also sing a song from a Broadway musical because just as the Saturday morning ritual was synagogue the Sunday morning ritual at our house was for my father to put on records of musicals and badly but happily sing along. This was the one day of the week we saw him for any length of time. (The wonder was that he'd never actually been to a Broadway musical in his life.) To this day my brother and I know the words to most of Lerner and Loewe, Rodgers and Hammerstein, Hart, Berlin, and Cole Porter. We were a family who loved song, from the synagogue to Broadway.

What was best about the Passover celebration each spring was that it combined all of these things—singing, performing, ritual, storytelling, and eating—among a huge animated gathering of family and friends. The table was set with china and silver reserved for the feast, all kinds of glassware and serving bowls we never saw any other time of the year. And, because this was a celebration of spring, of a life of slavery past, of a long journey, and a new life begun, we filled the bowls with mixtures of nuts, dates, figs, and sweet wine. One plate held the symbols of the season—a roasted shank of lamb, an egg, fresh greens—ancient symbols of renewal, that is, nothing that hadn't been seen as spring icons since prehistory.

Taking turns going round the table, singing and reciting (mostly in Hebrew that everyone could sing but only my grandfather, my siblings, and I understood), we told the story of the Jews' liberation from slavery in Egypt that each one of us was to take as our own story, as if each one of us had been a slave in Egypt and could know then the joy of a new free life. By accepting the story as one's own, one joined in a bond with all. That's what spring is for, reconnecting with the world and with one another. Through these rituals, from Mardi Gras to the Crucifixion to the Passover Seder, we reestablish our community, linking us one to another beyond blood and marriage, through a shared history, whether that history is real or not.

Mythology, as Ambrose Bierce put it, is "the body of a primitive people's beliefs concerning its origin, early history, heroes, deities and so forth, as distinguished from the true accounts which it invents later." He was more right than he knew. For instance, DNA research among a group of tribes in western Asia showed that despite all of them claiming to be descendants of Genghis Khan they weren't even related to one another. Yet they had at some point in time created a community and crafted a common history. They weren't true-to-life narratives. In the same way, there's no convincing archaeological evidence that the Jews or any other nation were held as slaves in Egypt, that there arose a Moses or after him a Canaanite-conquering Joshua. There's much archaeological evidence, in fact, to the contrary and much historical evidence that these stories were already ancient folktales by the time they were written down in the 7th century B.C. This is not news to biblical archaeologists any more than it should be news that no one named Noah actually sailed a ship full of male and female animals or that the world was not created in six days. They're myths made to forge a common consciousness. History is history. Faith is hope coupled with ideality. Spring was the time to renew faith, to celebrate the community's release from the constraints of darkness. The true and more meaningful story is that in the darkest times for Jews in Spain, Portugal, Russia, or Germany, in basements or behind closed doors, people still took it upon themselves to celebrate light's new dawning.

Some of the reason for this trip is to see things anew. I grew up very connected to the past. My grandparents married young; my parents married young and had children right away. Friends of mine knew only one or two of their grandparents. I knew my great-grandfather and three great-grandmothers and I remember the magical Passover in my great-grandfather's little apartment on Manhattan's Lower East Side, when we all sat together—great-grandfather, grandfather, father, and I—four generations at one table. And I saw my great-grandfather stand and dance, waving his handkerchief in the air, his gray beard aflutter. On this night, as young as I was, I understood the movement from darkness to light, from Polish shtetl to Allentown, Pennsylvania.

My brother and I are probably the last of my family who'll see four generations at a single table, and my brother, my sister, and I are the last of our family to even be able to recite all the prayers in Hebrew. But why should it be otherwise? Things change whether we want them to or not. That's what the Passover is about. The word "Easter" may have come from the Old English *eastre,* a festival of spring named for the Anglo-Saxon goddess of fertility, Eostre. Or it may have come from the Latin *alba,* "dawn," that became *eostrum* in Old High German and then English. The *Brittanica* doesn't seem certain. Either way, that's what this season allows: hopeful change, new life, light and time regained.

Here's my recommended reading for the Passover, Easter, Lupercalia. I read it at every Passover and, being away for this one, I have it with me and read it again:

Today, in this place, our only purpose is to reach the spring. At the moment we care about nothing else . . . in the morning, when it is still dark, we all look at the sky in the east to spot the first signs of a milder season, and the rising of the sun is commented on every day: today a little earlier than yesterday, today a little warmer than yesterday, in two months, in a month, the cold will call a truce and we will have one enemy less.

Today the sun rose bright and clear for the first time from the horizon of mud. It is a Polish sun, cold, white and distant, and only warms the skin, but when it dissolved the last mists a murmur ran through our colourless numbers, and when even I felt its lukewarmth through my clothes I understood how men can worship the sun.

PRIMO LEVI, *Survival in Auschwitz*

3

We Are All Biosphereans Now

April 11

13 HOURS, 2 MINUTES OF DAYLIGHT

West Virginia stands at the cusp of spring. Trees just budding in Clarksburg are near full bloom in Weston. Every mile south that I drive along this low plateau—Weston to Gem, Gem to Duck, Duck to Cotton—brings me further under the season's influence.

When I reach Charleston, West Virginia, I phone my laconic friend Jim Malusa in Tucson to let him know that I'm on my way, that I have meetings at Duke in Durham and will then begin traveling west to meet him. Back in January, when I called to ask him for the best time to see spring in the Arizona mountains, he told me in his very Zen-like way that whenever I arrived spring would be there.

"If it's summer at a thousand feet it'll still be spring at three thousand feet," he said. "If spring's gone at three thousand, we'll go to five thousand. And whenever you come, Sonya, the kids, and I will be glad to see you."

Now he tells me that winter brought good rains, spring arrived early, and summer might come in any day.

"Get here as soon as you can" is his uncharacteristic admonition.

"Okay," I say, "let me see what I can do."

My idea had been to explore a bit of the Appalachians and see some friends in North Carolina, but if I don't leave for the west until next weekend it will be two weeks before I'm in Tucson. I call the researchers at Duke to see if I can change my meeting date to tomorrow. They agree and so I head off to Durham.

The man I'm eager to meet is William Schlesinger. Dean of Duke's Nicholas School of Environment and Earth Sciences, Schlesinger's 2001 paper on the ability of forest soils to store carbon has been, since its publication, one of the most frequently cited publications in all of scientific literature, especially in global warming research. What made the paper possible is an extraordinary experiment that gives scientists a look at a forest of the future when the levels of carbon dioxide in the atmosphere will be 200 parts per million higher than they are today—a future that may arrive in anywhere from 40 to 100 years.

When plants photosynthesize they take molecules of carbon dioxide out of the atmosphere in order to create usable nutrients. Much of the carbon in those nutrients ends up in the soil. In the 1970s researchers found that plants exposed to high levels of CO_2 grew faster and bigger and that the soil absorbed the extra carbon. Based on this finding they hypothesized that increased atmospheric CO_2 might not be as much of a concern as they thought. Plants would suck it up and the soil would hold it in reserve. Schlesinger's work quickly quashed those hopes.

April 12
12 HOURS, 59 MINUTES OF DAYLIGHT

As it turns out, it's Schlesinger who meets me at Duke's 8,000-acre experimental forest. A tall, balding, gently robust man in a dark suit wearing large eyeglasses, he wastes no time getting us out to his futureworld in the woods.

"Everyone thought that if the trees and the soils were going to take up all the excess carbon dioxide we might reach a utopic state where we could still burn fossil fuels and just let stimulated plant growth take up the emissions. People said not to worry, the forests of the U.S. were taking up large quantities of emissions. I said, 'Hmm, what forests are these?'

"All of the research before ours had been done in greenhouses with pot-grown plants. This didn't apply to real nature or realistically include the competition between plants duking it out for nitrogen and phosphorus—'may the best tree win.' So we decided the only way was to do a big experiment outdoors. If we didn't find the uptake effects in these loblolly forests we wouldn't find it anywhere."

Loblolly pine forests cover much of the southeastern United States. The first trees to colonize abandoned farmland, they grow for 50 or 60 years and then, because their own seedlings can't grow beneath their shade, give way to oak and hickory trees. This section of the forest was planted in 1983. The canopy is dense and the trees close together. We're following a path that winds through the forest and that connects a dozen experimental plots, groves of trees that Schlesinger, his students, and scientists from around the world have been studying since 1993.

The farther we go into the forest, the louder is what sounds to me like the rush of escaping steam. After a few hundred yards we reach the sound's source. Encircling a group of pines some 30 meters in diameter stand several 60-foot-tall pipes of white PVC. A blower the size of a home's heat pump is at work.

"The blower pumps carbon dioxide, which comes out through holes in the pipes," says Schlesinger. "Enough CO_2 to add 200 parts per million to the ambient atmospheric CO_2 that was 360 parts per million when the experiment began and is now 380 parts per million."

This is Schlesinger's real-world, "may the best tree win" experiment—trees living in the future greenhouse. Three more such plots lay spread out through the forest along with another three identical plots into which no CO_2 is being pumped (although to keep the playing field level, there's a blower that blows but pumps only air). Schlesinger and his research team spend their time comparing the tree growth in these groves down to the last detail. They not only measure the production of pine needles, wood, and seeds, but they also tap the roots and the soils to analyze them for chemical changes.

We have to talk loudly over the sound of the blower and my first question had nothing at all to do with the trees:

"What would happen if I stood here breathing this air?" This is, after all, the air my grandchildren will be breathing.

Schlesinger smiles. "I don't think it would be a good idea, but eventually you'd adapt to it." In fact, it's probably no worse than the air in a crowded office, which, by Schlesinger's experimental results, should, if nothing else, be a good atmosphere for windowsill plants.

"In the initial years of the experiment," he says, "tree growth went up 20 to 25 percent. Within two years the production of cones and seeds went up 300 percent with three times as many seeds in the experimental plots than in the controls. But the big change was belowground, in the roots. There's been an 80 percent increase in roots. My interpretation is that when you add CO_2 the trees increase their rate of photosynthesis, which means they have all these carbohydrates but they don't have enough nitrogen and phosphorus or water so they take all this CO_2 and build roots."

Root systems put carbon into the soil but it doesn't just sit there. "Soil has its own respiratory system," says Schlesinger. Carbon storage increases between stems, forest floor, and soil, but at a certain point the soil can hold only so much. When spring rains soak the ground the interactions of microbes, minerals, and carbon re-create carbon dioxide. The soil exhales and releases CO_2 back into the air. Between plants and soil the forest does a lot of heavy breathing.

Before Schlesinger published his results many scientists were touting not only the forests' ability to nullify increases in CO_2, but also CO_2's ability to enhance the growth of plants and crops. But Schlesinger says that there appears to be a limit to tree growth from enhanced CO_2. (Other scientists have since found the same applies to crop plants.) Within the last year his forest's growth rate has begun to slow, producing more needles, roots, and seeds, but not bigger trees.

"At the present rate of CO_2 increase," Schlesinger says, "by 2050 the forests may be able to remove only 10 percent more carbon than they can now. If emissions continue at the present rate, that bit of increased storage will have little effect on rising levels of CO_2."

Duke's forest researchers keep track of everything that grows in the experimental plots and as it turns out, Schlesinger tells me, the big winners in this forest of the future appear to be the vines that crowd the understory. Since vines don't have to put any energy at all into producing wood or stems they thrive in the increased CO_2. The biggest winner of them all increased its growth rate by an extraordinary 70 percent. That big winner was poison ivy.

So if forests can survive increased levels of CO_2, can they also survive an altered climate? What Schlesinger and other Duke researchers tell me changes my basic assumption that warming itself will cause the most harm to plants, animals, and humans. The bigger and more immediate problems lie rather in the period of transition from one type of climate to another.

As the world warms over the next decades, melting glaciers, altered ocean currents, and wavering storm tracks will create erratic weather disturbances. Drought may set in where rain was once plentiful. Rains may increase where temperate conditions ruled. Sudden spells of warmth may be followed by just as sudden freezes. This more volatile atmosphere may engender more and more damaging heat waves, rains, hurricanes, thunderstorms, and tornadoes. All of these events will have heightened effects in the spring, when plants, crops, and humans depend upon a regimen of gradually increasing water, warmth, and light. Klemens, the herpetologist, has already seen, under the influence of early warm February rains, salamanders rush to their vernal pools to mate, only to find the weather suddenly frigid and deadly. With their breeding places so limited by sprawl, the loss of a couple of generations can threaten the survival of populations. The same will hold true for spring-hatching insects and species caught in midmigrations.

Full-grown trees are pretty tough. They can deal with the stress of gradual climate change and the fluctuations in temperature and moisture that come with it. But, like human children, trees are most sensitive during their first couple years of life. Seedlings' demands for warmth and water are critical. A mature forest will hang in there until a major event destroys it. However, a forest that began growing 50 years ago grew under conditions that, because of climate change, no longer exist. If a storm, drought, or fire should destroy that forest, what kind of forest, if any, will grow back?

For instance, the melting of polar ice has caused sea levels to rise a half centimeter a year for the last 50 years. If a storm fells a coastal forest that began growing when sea level was a half meter or so lower, seedlings will not find the same conditions that nursed the previous generation. The same trees will not grow in the now sea-saturated soil. Sawgrass marsh may grow in its place. Should storm surges wreck the marsh and flood the land, the marsh may never return.

These changes will happen on both larger and smaller scales. In New England, there's already evidence that as the present crop of sugar maple trees die, new sugar maples are not taking their place. As bark beetles acclimate to pine forests in the Northwest, they've begun infesting trees already stressed by recent years of drought. Should these trees die and burn, pine seedlings may not be able to regenerate under warmer and drier conditions.

The effects become magnified when birds come back in spring to an empty larder where there once was a marsh or forest full of food, when amphibians return to find their pools dried up, when polar bears in the Arctic find their ice pack gone, when places that had served species as home and hearth for thousands of years are destroyed, irretrievable.

As we leave the forest, Schlesinger invites me to attend a seminar he's giving this afternoon. I arrive early, watch the students file in, and once Schlesinger begins to speak see them engrossed by his effortless commingling of ideas and data.

"The organic carbon in soils and the CO_2 in the atmosphere are linked by plant photosynthesis," he says.

This, it seems to me, together with beauty and truth, is pretty much all they'll need to know.

April 13
13 HOURS, 1 MINUTE OF DAYLIGHT

Walking around Durham looking for a place to eat, I pass by the Duke campus, where students stroll bright lawns among distinguished old buildings, and I recall the fluidity with which, during my college years, time passed. Classes, music, marijuana, music, politics, music, seminars, cigarettes, sex, and books that I read as if each was a possibly canny proposition that I had to weigh carefully before being seduced, as I mostly was, by its argument, charm, or insolence.

Fortunately, hunger allays my drift into nostalgia. I buy a couple of slices of the local pizza (redolent of oregano but without a hint of Brooklyn), take out my map, and consider the long road west. Arizona lies 2,280 miles away. If the Chevy can handle it, I can cruise at 70, maybe 75, on the highways, get lost, give in to distractions, and still make it in time to see the desert spring. I might even drive a few hours tonight. I fold up the map, and when I see the young man behind the counter working the dough with a rolling pin, I refrain from recommending a different ratio of sauce to mozzarella, and step out into the humid dusk.

The car comes to life and we hit the highway. With my window and my radio the only contact with the world outside, I press on through the Gospel Belt and into the North Carolina mountains. In a motel outside Asheville (I've done a good three and a half hours) I turn on the television, undress, take out my well-worn wedding-gift flask, pour a drink, and watch the local news. Researchers, they say, have developed an equine vaccine for the West Nile Virus that last year spread into the Bluegrass State and infected nearly a hundred horses. Horses? I recall that last year West Nile killed 250 people in the Midwest alone.

Carried by mosquitoes and spread when they bite and draw blood, the virus was first identified in the West Nile district of Uganda in 1937 and hadn't even been seen outside the Middle East until 1996, when it appeared in Eastern Europe. An outbreak of a tropical virus so far north worried some epidemiologists, who saw in West Nile's spread a future in which global warming produced what one report called "an increase in the distribution and abundance of mosquito vector."

In the spring of 1999 the mosquito vector found to its liking the northeastern United States' hottest, driest spring in 100 years and West Nile appeared in, of all places, New York City. The mosquitoes infected thirsty birds that came to drink in small stagnant pools of water. The birds flew off to spread the virus or die. When rains broke the heat and drought, the mosquitoes moved from the ponds and took the virus with them. By summer's end the epidemic had infected 8,000 people and killed 7 of them. People became so worried about mosquitoes that they nearly forgot their fear of ticks and Lyme disease.

In the next year West Nile spread throughout the East Coast. Health officials reintroduced spraying for mosquitoes, and they warned people away from water where the insects might breed. By spring 2003, the mosquitoes had moved west, bringing the virus to the drought-stricken Plains states—Nebraska, Colorado, and the Dakotas.

Is the spread of the virus a result of climate change? The World Health Organization thinks so and warns that the "mosquito vector" may soon be finding more of the northern hemisphere to its liking.

But then, as my 16-year-old daughter put it, "When anything happens in school we just say, 'It must be global warming.'" Fail a test? Lose a boyfriend? Global warming. In the morning paper I pick up in the motel's lobby, I read that global warming may prove detrimental to the taste of fine Kentucky whiskey. Since the final maturing of quality bourbon depends upon the seasonal warming and cooling of the barrels in which it's aged—the barrels' expansions and contractions imparting both color and flavor to the bourbon inside—the proper range of temperatures is critical. You couldn't, for instance, make Kentucky bourbon in Georgia. But if the Kentucky climate warms too much more that's exactly what they'll be trying to do. "Oh, for a beaker full of the warm south," wrote Keats. I'm glad I'm stuck on Tennessee sour mash.

I cross through the long-suffering stands of Smoky Mountain firs and down through the regrettable haze of the hardwood forest—regrettable because winds that move Midwest pollution eastward often bring to this extraordinary national park ozone levels as high as in Atlanta, nearly as high as Los Angeles. Since 1948 visibility in the Smokies has been reduced by 40 percent in winter and by 80 percent in summer. The vista of greening beech and maple has a dusky blue cast and the sunlight seems to be fighting its way through to the forest. Still, when I take a break and a walk, I can't help but feel the understory's restorative exhalations.

Sprouts of green make me decide to stop at a nearby diner in the hope of being served a salad of fresh-picked ramps, but I'm told that pickers are banned from the national park and outside the park the pickings have been slim. The ramp, a wild leek, has the sweet tang of a wild onion, to which it's related, but the musty odor of garlic stored in old socks. As a fellow at the diner offered when he heard my request: "You don't want to eat ramps unless your wife eats them or you'll get thrown out of the house. I once ate them in the morning and they threw me out of school."

Still, since it is the first edible green to come up in the spring, its arrival has long been a cause for celebration. Local festivals from the little Cosby, Tennessee, ramp festival to the grand Elkins City, West Virginia, international (both still a couple of weeks away) go through some 3,000 pounds of ramps a year. That, apparently, may be too much, threatening the survival of this trenchant taste of spring. I settle for a burger and a salad with garlic dressing.

As I drive the long stretch toward Memphis, the Chevy and I feel the ingrained lure of the land beyond the Mississippi and decide for a time to keep to the highway. I choose quiet over the ugly harangues of radio evangelists. But the quiet is never silent. Without the beat of my heart to attend to as I drive, I lose touch with myself. When I'm driving at night I feel insensate, as if I could be anywhere. This may be what you get when you borrow a car from a Heidegger scholar: I am, as Miles translates his mentor, trying "to capture the imminence of the exsistential or forward-reaching future." When my eyes blur so that I can't focus on the highway signs, I stop, take a room, and fall asleep still feeling the vibrations of the car and the highway.

Tonight I pass a small grocery store when I pull off the highway, but instead of stopping I decide that since the muscles in my legs feel like they're withering, I'll check into the motel and walk back. There's nothing else along the narrow blacktop road and if the vaporous glow from the single light in the store's parking lot goes out the store itself would vanish in the dark. From the unmowed roadside grasses the panicky stridulations of cicadas seem to set the night's mist to resonating and I feel as if I'm walking through a haze of sound.

This is the spring of the 17-year locust, as the cicada is known, although the insect isn't a locust at all. They've spent the last 17 years as grubs buried several inches in the ground surviving on the nourishment they can suck from root sap. When the ground temperature reaches 64 degrees (they're very exacting) they tunnel to the surface, inch up the nearest tree or wall, and do what every insect does in spring—metamorphose, come out of their shells, take wing, sing, and mate.

Of all the cicadas—and there are 7-year, 11-year, and a dozen more other-year cicadas—the 17-year appear in the greatest numbers. Up to a million an acre. Soon enough they'll seem to be everywhere, dark-bodied flies with orange-red eyes and orange-red-veined wings. When the males set their sights on mating and sing—and only the males sing—the chorus can be ferocious. They invade in such hordes that all the birds, squirrels, dogs, and even people who feast on them (cicada cookies, cicada cakes) hardly make a dent in the population. Some cicada mathematicians have even figured out that by returning every 17 years these cicadas stand the least chance of coming up out of the ground at the same time as other periodical cicadas and thereby increase their chances of survival.

Their springtime, however, is brief. In the few days they have they mate, place their eggs in tiny boreholes in trees, and die. The young hatch a few weeks later and crawl, mostly unnoticed although still by the millions, down the trees to vanish into the ground. Sex, death, new life; cicada, salmon, mosquito; the odd, the elegant, the changeable—nature's the maddest collection of them all.

I reach the little supermarket and the fluorescent hyper-reality is a shock to my tired eyes. I assume I look like hell to the pleasant-faced young clerk, so I smile as best I can and he, properly, I think, considering my unshaven and glassy-eyed appearance, doesn't smile back. Aside from the nuts and raisins I've been munching on while I drive I haven't eaten since breakfast and it's now nearly nine. My question, as I wander blankly through the store, is whether to buy a box of cookies, a bag of pretzels, have a drink, and go to bed, or buy some cold cuts and bread, make a sandwich, have a drink and go to bed. But inspiration strikes as I pass the frozen food case. Home cooking! I find a box bearing an appetizing photo of burritos, the label reads "All Natural," and I grab it. I buy some more nuts for tomorrow's drive, a banana for breakfast, a bottle of water, and I'm set. The clerk swaggers a bit as he comes over to check me out and I make a stab at conversation.

"What time are you open to?"

"Ten."

I nod.

"You have the newspaper?"

"All sold."

"Well, thanks. Good night."

The motel's glow seems far off and the walk to reach it seems long. I might be too tired to stick this burrito in the room's microwave. Quiet town. I'll have to look at the map and see where I am. Somewhere between Oklahoma City and Amarillo. Without my heartbeat, I use the cicadas' vibrations to provide me with a connection to time as I walk. Am I missing my old heart? Perhaps this happens. A friend of mine once called me in a state of emotional panic and when I asked her why she stopped taking her medication she told me she missed her depression. Larry McMurtry, in his autobiographical book *Walter Benjamin at the Dairy Queen,* writes that he feels as if he's had two lives, one before and one after his open-heart surgery. McMurtry, with some prescience, blamed his altered sensibility on the heart-lung machine. I suppose I could say the same but I'm not certain I have a new life yet. I'd say I'm in between lives. Not unlike the cicadas arising to complete one existence and begin another.

April 16
13 HOURS, 7 MINUTES OF DAYLIGHT

The car's running fine, better than fine. He roars a bit going from 50 to 70, like a weight lifter powering a clean and jerk, but once at 70, momentum alone pushes him to 75.

The plains rise and near the Oklahoma border snow, sleet, and ice blow in on slashing winds across the open lands, where red wildflowers already bloom. The fields, where seedlings have just begun to rise, nearly vanish in the manifesting fog, and soon I can make out only dim images of grazing cattle and bobbing goose-necked oil pumps. Along the road, lights flash at little combination quick-mart casinos owned by Indians.

I turn on the radio.

"God calls it murder!"

I turn off the radio.

I stop at a restaurant and have chicken-fried steak. God seems appeased. The storm passes. I decide to call Jim and let him know I'll be there Saturday.

"It's 85 here," Jim tells me. "The ocotillo is flowering."

"Well, I'm in Oklahoma. And it's fucking snowing."

"Get here by Sunday. Abuela's invited you for her paella."

"I'll be there Saturday."

I drive off into a brightening but cold, humid dusk. The winds pick up from the southwest. Ducks on the fields up and fly with it.

Out on the high Texas plains it turns dry and warm. The land's cracked and disjointed, full of potholes, gullies, and ravines, but it soon turns so flat that there seems to be no horizon. Blue clouds lying low in the immeasurable distance look like silhouettes of distant islands. The setting sun behind them limns their shorelines in gold. I've seen such islands. It was the Andaman Sea, off the western coast of the Thai peninsula, and for a moment the sky becomes another place altogether. If a traveler doesn't allow himself to hazard these kinds of correlations, what's the point? Reality's always only a flick of a switch away. One flick and I hear the sounds of artillery in the Middle East. Another and I hear that a nearby high school has to have three different proms—White, Black, and Hispanic—in order to keep fights from breaking out. I hear that at its present rate of decline due to drought, Lake Powell, the great reservoir behind the Glen Canyon dam on the Colorado River, has only enough water to last three more years. What more can I want to know? If you had a gallon of water for every politician's promise that Lake Powell would solve the West's water problems, the damn thing would be overflowing its banks.

It's cold on the Texas plains but the dead flat vastness of the fields gives them a mellow dignity. Spring-born colts lean against their grazing mothers, nosing to nurse. Then, over a rise, at Groom, Texas, it appears: "The Largest Cross in the Western Hemisphere," a 190-foot-tall monument—to Jesus? to Christianity? to Everything's Bigger in Texas Christianity?—with a white metallic gleam that stands in the middle of nowhere. An unofficial guide I find (*Route 66 Unraveled* by Howard R. Sauertieg) says, "As pious erections go, the Great Cross of Groom is almost Heliogabalic in its aspirations." (Now this intrigues me and I find that Heliogabalus was a 3rd-century Roman, a priest of the Sun god, who was also known as Elagabalus, Varius Avitus Bassianus, and who, as Marcus Aurelius Antoninus, reigned as emperor between 218 and 222. A preposterous figure known for his gluttony and sexual weirdness, he had his chariot drawn by a team of naked women.) As I drive by the Texas-size monument, I see a busload of Japanese tourists standing in the parking lot looking up and taking photos of what looks to me like an overreaching, Heliogabalic telephone pole.

I head for the hills of New Mexico.

April 17
13 HOURS, 5 MINUTES OF DAYLIGHT

I've begun to recognize when I'm in Indian country. In the three-card monte the U.S. government played with Indian land, the Indians ended up holding some of the most arid, rocky, busted, and barren places in the West. Somehow, with all the open range in this endlessly expansive country, we found it necessary to have it all. That Indians couldn't make a go of it on these paltry reservations should surprise no one. What amazes me, though, as I drive through these dusty outposts, is that while we send good men and money after bad to shore up nations abroad we're still able to ignore the nations struggling to survive within our own borders, nations within a stone's throw of the World's Largest Cross.

Perhaps I'm just prickly with the heat. Perhaps it's the billboard on the talus hillside outside Gallup, New Mexico, that bears some badly executed drawings of tepees and tomahawks and reads, "Photograph the Past." (Would we could.) Or maybe it's just that in the desert every kind of conflict seems to have a heightened intensity. Even with all their marvelous adaptations to living in arid lands, a plant's or an animal's survival is rarely assured. Humans—at least those who pushed the Native Americans to the plagued peripheries, those who live with little regard for how and where—assume that survival is the least of what they have coming to them. Look, for instance, at Las Vegas, or Phoenix. If these sprawling cities speak of anything it's of grandiose human presumption; that is to say, inferences drawn from beliefs based on unproven assumptions (the major one being that the supply of water is endless), ignorance amplified by overconfidence (once again, that the supply of water is endless), and just plain disrespect for the nature of the world around them (that the water is theirs). Which is to say that two hours from Tucson I'm stuck in Phoenix, in traffic mired in ovenlike heat, and the Chevy, a purring giant for all of these thousands of miles, begins to buck and balk.

I pull into a gas station, put the Dick in neutral, and gun the engine until the temperature needle drops back into the normal range. I let him idle for a time, turn him off, and get out to buy myself some water. When I come back, a broad, muscular guy in a red polo filling his red Mustang with gas comes over.

"Great car," he says.

"Thanks. Getting a little hot in this traffic."

"This?" he says, waving his hand in the direction of the jammed highway. "Welcome to Phoenix, Los Angeles East, the postmillennial metropolis. Where are you from?"

"Brooklyn, New York," I tell him.

"Old World," he says. "How'd the car do?"

"Okay until now."

"What kind of mileage?"

"About twenty on the highway."

I'm pouring sweat. Somehow, he seems cool. Postmillennial metabolism.

"What's my best way out of here?" I ask.

"Look, just take your next exit off the highway, take that road south, go about twenty miles, and you'll be able to connect back up with the highway outside the city."

"This traffic's worse than it is in New York."

"Highways are all we have. Everybody's on them all the time."

"Thanks," I say.

He nods and heads back to his Mustang. I stand and drink my water. Why "postmillennial"? I believe it's clear by now that when humans enter an isolated system we try to control its natural entropy. Open space, like unmeasured time, has no evident borders, no "yours" or "mine." Why would I think that people moving out here to the desert would all become Frank Lloyd Wrights and evolve a new architectural ethic to match their environment? Wouldn't they create the same morass of highways and fill it with cars just as we did everywhere else? Just as the first Europeans believed in the enduring vastness of the continent's forests, so the pioneers believed in the capaciousness of its western mountains and deserts. Perhaps a "postmillennial metropolis" is one that, imagining itself loose from the past, has chosen to ignore the lessons learned during the previous millennium. Which is as much to say, as British comic Peter Cook said, "I have learned from my mistakes and given the chance I could repeat them exactly."

April 17
13 HOURS, 2 MINUTES OF DAYLIGHT

Botanist Jim Malusa works at the University of Arizona but spends most of his time out in the mountainous desert mapping the vegetation for the National Park Service. He heads out alone with several gallons of water on his back and hikes for a week at a time, covering the desert quadrant by quadrant, recording the mix of plant species. When he returns to his office he enters his records into a database that then places the plants on a digitized map with which the U.S. Fish and Wildlife Service, or anyone with a computer, can scan the desert in three dimensions. In this way they can track changes in the desert environment and its habitats in virtual time and space, soaring low over terrain most of them will never experience. Even fewer will know the special kind of strengths, physical and spiritual, it takes to make the demanding, solitary journeys that brought the map to life. Malusa is a special kind of desert denizen. Sentient and exacting or bleakly irreverent, he's part adventurer, part monk.

"I've hiked alone ever since I can remember," he tells me. "And unless it's good company, I'd still rather hike alone."

Good company includes his wife, Sonya, their two wildly curious, dark-eyed kids, Rudy and Rosita, and a few select friends. I have the honor of being considered good company and one of the family. Without having to ask I can set out my sleeping bag on the floor in Jim's home office and make myself at ease, which is exactly what I do when I arrive at their house in Tucson. No one's there but I let myself in and, exhausted to near delirium from the week's drive, I try to make up for all my lost Eastwoods. I awaken to the sounds of Sonya shushing the kids to let me sleep, get up, and find out that if I'm interested I can go along with them to tonight's spring fair at the kids' public school, an offer that I can't refuse. So with Rosita and Rudy taking turns riding on my or Jim's shoulders we take a warm night's walk to the fair, where Hispanic, African American, Asian, and White kids and their parents are together enjoying hot dogs, bake sale cookies, cotton candy, and carnival rides. I find the tensions of the drive dissipating. I've made it west and take this pleasant children's Bacchanalia as a good omen. (Bacchus, I'm sure, would have loved cotton candy as much as wine and flowers.)

The next night brings the anticipated visit to Sonya's mom's house for paella. Abuela, as the kids and family call her, was born in Spain and along with her family suffered there through the Spanish Civil War years. Paella, for her, is as much homage to her native land as it is food. A kitchen stove won't do. Paella by Abuela requires a wood fire that she herself has to build and stoke, splint by splint, in order to maintain the proper temperature, the right ratio of heat to smoke, while she cooks, in methodical sequence and in nothing but a proper *paellera,* the chicken, rice, shrimp, clams, squid, saffron, and vegetables. In the end she conceives a toothsome feast, timed to be served with the desert sunset. Twilight is suffused with the scent of orange blossoms, saffron, and wood smoke and fêted with a good deal of Rioja. Lively conversations among Abuela's friends and family switch back and forth between Spanish and English, and return often to remembrances of the war that 70 years ago yoked their homeland to fascism and sent the freedom-loving into hiding or exile until the death of Franco and the spring of Spanish democracy. The evening ends with a general rousing of sleeping children and embraces among new friends.

April 19
13 HOURS, 6 MINUTES OF DAYLIGHT

The next morning Jim and I pack our gear and head out east from Tucson in his aged and properly beat up Jeep, top down because there's no top to put up. He's agreed to take me to see the desert spring although, mordant coyote that he is, he says that he himself finds far more excitement in the desert's late summer, when storms can strike and in a flash fill up a dry canyon. Spring, he admits, has new growth, cactus blossoms, birds, and comes as a relief from winter, but then winter in the desert is hardly noticeable.

"Just show me spring," I say.

"Okay." He smiles. "If that's what you're here for, we'll find spring."

Getting out of Tucson's metro area is not what it used to be. Development has reached into all the surrounding foothills; the only thing keeping it from climbing higher and perching on the mountainsides is that 50 or so years ago the federal government took control of what was then remote desert. So sprawl spills across the basin, clusters of retirement developments conjoined at shopping center nexuses.

Beyond the sprawl lies the Sonoran Desert, a desiccated plain of rock that stretches from Arizona to Mexico, inhabitable by only the hardiest of species: mesquite, paloverde, cactus, birds, rodents, reptiles, pronghorns, and coyotes, each with its life-sustaining adaptations. Up in the hills the Keith Haring–esque silhouettes of giant saguaro cactus dominate the landscape. These inscrutable giants can grow to 35 feet, weigh 10 tons, and live 200 years. Their long lives begin humbly: seeds must sprout in the shade and develop under the protection of less majestic species such as prickly pears, hook-spined chollas, and barrel cactuses. After 10 years, a young saguaro stands barely an inch high and another 70 years will pass before its woody skeleton grows strong enough to support branching arms. Only then will they begin to produce flares of large white flowers and jam-worthy sweet red fruits.

Being from the East, where most of my hiking is done in dense forests of oak, pine, and hemlock, the desert panorama takes some getting used to. In every direction a range of mountains truncates the vista, massive arcs of ancient rock that rise as high as 18,000 feet from the 20-million-acre sea of desert that surrounds them. But unlike the Rockies, Cascades, or Appalachians, these ranges are not continuous. They stand apart from one another like islands and form an archipelago of mountains that stretches from the Sierra Madre Occidental in Mexico north through the Huachuca, Patagonia, Catalina, Rincón, and Santa Rita ranges, to the Galiuros northeast of Tucson.

Looking toward every horizon and seeing no shade in sight, my mind thinks, "wasteland."

It wasn't always this way. Before the climate in the Southwest began to change some 15,000 years ago, dense forests stretched from the Rockies down to the Sierra Madres. The animals that lived in these forests—wolves, black bears, jaguars, squirrels, birds, reptiles, and insects—moved freely north and south along a lush and wooded corridor. Tree and plant species readily spread from one place to another. But with the end of the ice age, temperatures rose. Precipitation dropped. First, the forests farthest south disappeared. The land turned to desert and the desert spread north, forcing the forest life into retreat, seeking cooler, moister climes where it might survive. Many species found refuge in the mountains, moving higher as the desert formed all around them. Those that did survive found themselves stranded, cut off by wide arid gulfs from others of their kind, imprisoned in their places of refuge. After thousands of years of such isolation the result is that, as Darwin reported from the Galápagos, "different islands to a considerable extent are inhabited by a different set of beings."

But the islands of the Galápagos at least share the same equatorial climate. In this sky island archipelago, as it's known, every thousand feet up that you climb takes you into a climate that's more like one some 500 miles north—a few degrees cooler and with a few inches more rain or snow. So up in the Galiuros we're among native Arizona saguaro cactus at 2,000 feet, grasslands at 3,000 feet, juniper at 4,000, oak at 5,000, and Rocky Mountain pine forest at 6,000. The Douglas fir forests at 7,000 feet resemble those of British Columbia. Here, as Malusa told me, it may be spring at 2,000 feet while at 6,000 feet it's still winter.

After a long dusty run in the Jeep we arrive at a place Jim prefers I call Red Rock Canyon to protect its serenity. We put on our packs and head out. Sometimes the trail is recognizable—a winding footpath worn into the hard desert rock. Sometimes it's even marked by a pile of stones. But most of the time we're either bushwhacking through tangles of scrub or scrambling with our full packs over boulders the size of VWs that fill the streambed between the narrow canyon walls.

These walls, steep, chiseled, and orange-red, rise 100 feet above us. When we look up, we can occasionally catch a glimpse of the sunlit mountain that rises another 3,000 feet above the canyon.

A vivid but motley assortment of trees and plants surrounds us. Few things grow in groves or thickets for down here, with water and good soil so scarce, it's each plant for itself—find a niche and make the best of it. For something to be where it is it has to outcompete something else. In places where water flows year-round, alders and sycamores line the stream bank. In their shadows the undergrowth stands dense and barely passable. Just above our heads, however, agaves and columbine sprout from the sun-drenched canyon walls. Twisted conifers cling to the rock in shaded alcoves. On opposite, narrow ledges of riverbank, a gnarled black walnut faces a lone aspen.

This specificity of habitat—the individual that, even separated from others of its kind, finds its own place in the world, species be damned—fascinates Jim.

"It's the way I look at the world. I see things and always ask myself why they are where they are. And there's almost always an answer."

Geography is destiny.

The pronghorn antelope, for example, adapted remarkably to the desert. They go long periods without water, depending mostly on the moisture in the cactuses and grasses they eat. With their red-brown and white coats, black manes, and black-masked faces they would not look out of place on the African plains or in a zodiacal pantheon. Some 35 million may have grazed the American West before Europeans arrived and hunted them nearly to extinction. These days, however, as drought becomes more frequent, the few places with dependable water often lie across roads and behind fences. This puts the pronghorn at a particular disadvantage. While out in the open desert a pronghorn can run 45 miles an hour, groups have been known to die of thirst or starvation rather than leap a low fence to get to food or water.

Where the canyon widens and the sun penetrates, our only water lies down in rock crevices. When we find one of these deep pools cold and clear enough to see it alive with small desert fishes, we pump some water through a filter and fill our bottles. Hot and tired, at one small pool I undress and wade in for a quick dip but can bear the frigid water only a few seconds. And once, while dressed, I fall into a pool up to my waist but the air is so dry that I'm barely wet long enough to cool off.

That night, we lay out our sleeping bags next to the stream and build a small fire. Jim cooks some beans that we have with cheese and tortillas and afterward we roll ourselves cigarettes, share drinks from my flask, and watch the constellations pass across the narrow window of sky above us. Tree frog calls echo through the canyon like the knocking of woodpeckers. Lying beneath this star and rock canopy, as if on a hammock swinging somewhere between Earth and Heaven, might seem like a good time to think things over. I try to wonder why I am where I am. But thoughts apparently have no solidity here. They arise but, like the sparks that try taking leave of the fire, quickly vanish. Once more I'm in the dark and I don't hear my heart. Once again I think that my journey's finally begun but I've thought this before. How many damn beginnings can a journey have? At some point you have to accept that, for better or worse, you're on the road. If each dawn's a crackerjack new day, each day the first day of the rest of your ever-commencing life, there's no way you'll appreciate the journey. Spring is here. I'll let my heart go romancing.

Overnight the temperature in the canyon drops from 80 to 40. Jim tells me that some mornings out in the desert the temperature gradient is so sharp that he's measured a 20-degree difference between the air around his feet and the air around his head.

We blow last night's coals up into another fire, make coffee, share oranges and the rest of last night's burritos, pack up, and head out. After investigating a blind canyon that we might have mistakenly thought to be our way out had Jim not carried a good topographical map, we begin to climb. The going at first is fine. In the lower canyon there's still soil beneath our feet and shade above our heads. But soon we come out into the full afternoon sun. For a few hundred feet we're surrounded by saguaros, but at 3,000 feet we pass through their narrow bandwidth and with 2,000 feet more yet to climb, the trail becomes rocky, steep, and overgrown with thorny shrubs of catclaw that tear at our clothes and draw blood from our legs and arms. I recall Ed Abbey writing that you haven't hiked the desert unless you come out bloody.

This south-facing slope is far too sunstruck to support any trees at all. I notice that I'm not perspiring, the dry air drawing off the moisture as quickly as it comes. I also realize that unlike the last time I hiked with Jim, last spring before my surgery, I feel no pounding in my chest and, although breathing hard, feel no weariness. My heart's floating like a butterfly and I find myself even smiling as we climb switchback by switchback to the crest of the ridge. We stop and sit beneath what low shade we can find.

"You okay?" Jim asks.

"Yeah," I say. "I'm okay. I feel good."

We drink and look back to have a last view at the canyon out of which we climbed. But it's gone. What had seemed to me a vast subterranean Jules Verne world has been swallowed up by the sprawling mountain vista.

Picking up again we can look north over our coming down-hill climb, a gentler slope brightened by blossoming ocotillo and just-fruiting chollo, with a meadow beyond, covered with yellow mustard flowers. But beyond that lies a rugged plain of mesquite that extends to the foothills of yet another mountain island that rises from another sea of desert. As we drive back at dusk the saguaros stand in growing shadows, golems waiting to be animated by moonlight.

The survival of desert plants depends upon the spring arrival of hungry nectar seekers. For instance, hummingbirds on their spring migrations track the ocotillo bloom. Long-nosed bats move north from Mexico following the bloom of columnar cactuses. White-winged doves follow the saguaro blossoms. Along with sweets-seeking bees and butterflies, all of these species make it on their long journeys north only by being there on time for the next bloom. To take full advantage of this migration and not use up their own slim reserves of water and nutrients, cacti don't bother growing leaves and instead put all their energy into flowering briefly each day (or night), but doing so with displays of large and luscious-colored blossoms that can't fail to grab a nectar eater's attention. Each time a bird or insect goes poking into a blossom for nectar it's dusted with pollen. That pollen falls onto the pistil of the next flower they bother. The principle behind this spring rite of food and fertility is neat: In gathering the food that fuels their migrations the nectar gatherers ensure the survival of the plants that provide them with the food they need along their route.

In 1993, bat ecologist Ted Fleming called this migratory path a "nectar corridor." Its importance to the ecology of the Southwest can be seen in the exacting detail in which it's been studied. The ocotillo will tell you how far spring's advanced. Also known as coachwhip, flamingsword, Jacob's staff, and candlewood, ocotillo stems—and the plant is all stems—grow in dense springs up to 20 feet tall that look like stands of waving sea grass. In deference to their environment, their leaves appear only after it rains and then drop when the soil dries. But the red flowers that line their stems like match flames appear with annual regularity. The wave of blooming sweeps from south to north across the desert lands. When the ocotillo is through blooming, the desert spring has ended.

Researchers at the Sonora Desert Museum outside of Tucson investigated a dogleg-shaped section of the nectar corridor that runs from just south of Guadalajara, Mexico, where the corridor is at its narrowest, to the Sonoran Desert on the northern Arizona border, a straight distance of 1,660 kilometers bordered by the Sea of Cortez on the west and the Sierra Madre Occidental on the east. They tracked the migrations of long-nosed bats, humming-birds, white-winged doves, and butterflies through mountain forests and two deserts.

In the spring the bats follow the blooming of cacti north, fuel-ing up on flower nectar along the way. To find out just how dependent the bats are on the flowers, the museum scientists worked out one of those word problems you hated in high school math. It goes like this: "A colony of 100,000 long-nosed bats flies north from their cave in Mexico 1,200 kilometers to Arizona. They fly at a rate of 50 kilometers per hour. If they gain 6 grams of fat before they leave and use 3 grams of fat to travel each 283 kilometers, and each visit to a flower provides them with about half a kilojoule of energy (each gram of fat provides 39 kilojoules of energy), how many flowers will they have to stop at along the way?" Please show your work.

The answer? Each bat needs to visit about 227 cactus flowers. Which means "that assuming each flower is visited by two or three bats it will take at least 10 million cactus flowers to fuel the spring migration" of this one colony of bats. How long has this relationship been going on? Bats and cacti both evolved some 40 million years ago. By now, without the bats much of the cacti along their route would disappear. But just as the cactus extended their range north when the southern deserts first formed, so they'll continue to do as the Earth warms and the deserts move north. Will the bats and birds follow?

The birds can do only what they've always done: respond to the changing light of spring and begin their migrations at the same time each year. Cactus flowers, however, bloom in response to temperature and moisture. If, in a warming world, spring temperatures come sooner, the flowers may have already bloomed and gone before the birds arrive. Thus, the timing of events that took thousands of years to synchronize gets altered in too short a period of time for biological clocks to adjust. Imagine that the train you take to work begins to come earlier and earlier each day but your alarm continues to wake you at the same time each morning. As the world quickly warms, the migrating species of spring will find themselves in just such a time warp. Without enough pollen and nectar to refuel they'll be too weak to go on to nest and breed. Over time, if there is time, adjustments might be made. Those bats or birds (and their offspring) that somehow began their journeys north with an extra gram or two of fat reserves might survive. But adapting to such a change would most likely take generations.

The present state of drought is not unprecedented. As the tree-ring scientists, called dendrochronologists, discovered, the fairly sudden appearance of long-lasting droughts, droughts lasting from a few to a hundred years, are the norm in the Southwest. The aftermath of such droughts can be devastating to humans as well as plants and animals all already living on the edge. When farming becomes impossible, topsoil blows away. Even rains that break the drought can bring ruin. In 1993 in the Four Corners area of the United States, where Nevada, New Mexico, Arizona, and Utah meet, heavy snows and rainfall followed several years of drought. Plants and animals returned in great numbers, among them deer mice that increased their numbers tenfold in a year. While holed up in close quarters during the drought they became carriers of a hantavirus, a virus related to those that cause hemorrhagic fevers such as Ebola. When the mice returned, people who came in contact with them or their leavings were stricken. Fever, chills, and aches led to fluid buildup in the lungs and in many cases, in only a few days' time, death. This was the first outbreak of hantavirus ever recorded in the United States.

But the virus may have struck farther south. Mexican epidemiologist Rodolpho Acuna-Soto has pretty much proven that in 16th-century Mexico, long-term drought followed by heavy rains caused outbreaks of a hemorrhagic fever that the Aztecs called *cocolitzli,* which killed millions of native Mexicans and completed the devastation of the Aztec kingdom. Drought and virus, submits Acuna-Soto, may have also been the cause of the seemingly mysterious disappearance of the Mayan cities some six centuries earlier.

By now the desert has become less of a mystery. In what once appeared to me to be wasteland, I can now distinguish lush and brilliantly appointed accommodations.

April 23
13 HOURS, 13 MINUTES OF DAYLIGHT

There is nothing like being out in the field with a good birder, and Malusa is nothing like a good birder. While he can identify a plant at a thousand paces, when it comes to birds his taxonomic powers weaken considerably. Perhaps his quest for "why things are where they are" becomes more difficult when the things have wings. In any case, I've known scientists who studied fish anatomy who couldn't identify a species with its skin on. So Jim recommends that I hook up with his friend Stephen Hopp, a University of Arizona ornithologist who is taking his students on their spring birding trip into the Santa Rita Mountains south of Tucson, just some 35 miles from Nogales and the Mexican border.

While Jim's approach to the world is, like Jim himself, angular and rascally, Hopp, with his full beard and professorial standing, is direct and droll. Both are clever, quick, and loathe letting work become more work than play. Being in the field with an ornithologist such as Hopp is like being at a blind tasting with an oenophile, in Hopp's case, a connoisseur not of taste but sound.

Song fills the birder's spring world. With so many species, birders learn to recognize the basic birdsongs by mnemonics that among the cognoscenti seem to have a universal currency. I say "teakettle teakettle teakettle" and you say "Carolina wren." I say "wheezy, wheezy, wheezy" and you say "black-and-white warbler." Had I said "zay zay zay zoo zee" you would have said "black-throated green warbler." "Creep-creep" without question is a great crested flycatcher.

But there's more.

"That blue-headed vireo," Hopp says to his students, "is its song coarse or sweet?"

They screw up their faces and listen hard.

"That—you hear it? Is it a call? Or is it a song?"

"Song," say half the group at the same time the rest say, "Call."

"Three peeps. You hear it? What is it?"

This time no one chances a response.

"It's going to be our bird of the day!" he says.

And it turns out to be nothing less than the elusive northern beardless tyrannulet.

At the Santa Rita's research station this first day out, to prove to Hopp my gameness for birds I crawl out of my sleeping bag at 5 in the morning to be there at "first light," the time when Hopp told his dozen or so students to be up and ready. I'd made my first easy bird identification during the night. "Grog, grog" was a garrulous raven who perched in the branches over my head and croaked like a beery sailor whenever it appeared to him that I might be taking sleep for granted. At 5:30 Hopp's young crew, in baseball caps with binoculars dangling from their necks, all gather. First light comes and goes, however, before Hopp appears, having, he explains, exercised his professorial field trip prerogative to sleep in.

"This is not a classroom," he says, grinning and obviously well rested.

Far from it. It's a mountaintop research station and preserve that overlooks the sprawling Green Valley. From our vantage, steep slopes descend through open desert rangeland tousled with twisted scrub, perfect shelter for wary birds. With the coming of spring these mountains fill with birds: those that have spent the winter and are preparing for their long flights north along with migrants from Mexico and South America that will spend the summer. All join the resident ravens, woodpeckers, and jays. Light and food are plentiful. The nectar-rich ocotillo is in full bloom. There's desert honeysuckle and primrose, lupine, agave, oaks, and junipers, lush habitat by any standards.

The sun at midmorning makes it feel like midday, but birdwatching is a halting walk in the woods. We walk and listen, and then Hopp plays master of ceremonies, giving his student contestants a couple of minutes to discuss it among themselves and name that bird.

I ask him how the birds find their way but the answer is that it depends.

"We used to think it was just a single migration," says Hopp. "But now we know that there are many migrations."

All birds respond to the changing spring light. Those with long journeys to make, for instance, will begin to put on weight. Their timings will differ based on the length of their migrations as well as the land over which they'll fly. The idea is to arrive at their various way stations when food is most available.

Navigation methods differ as well. Smell, light, landmarks, magnetic fields—all have been suggested for the way birds seem to fly unerringly toward their destinations. But it may also be that they use a combination of methods. For instance, those that rely on magnetic field navigation might also adjust their route each day by taking a bearing from the direction of the sunset. And sometimes, bird flights are not so unerring. Some, perhaps diverted by weather or winds, have been known to head in the wrong direction altogether. This may not always end badly. Should an off-course migrant find food and nesting spots along the way it might well survive to make the same mistake the next year and perhaps even colonize new territory.

Our bird list quickly grows: warbling vireo and a magnificent, iridescent-breasted hummingbird; yellow-rumped warbler and yellow-tailed warbler that will soon head off on a more-than-a-thousand-mile flight to Alaska. Goldfinch, Arizona woodpecker, and several nutty acorn woodpeckers that had drilled holes in the cabin eaves and used them to store kibble stolen from the field station's watch dog. A Townsend's warbler is on its way to the Pacific Northwest. From its winter haven in the Sea of Cortez comes an ash-throated flycatcher. We see a Wilson's warbler and a Lincoln sparrow that next week will be in Canada. And on a climb up a ridge thick with ocotillo we see a rattlesnake curled between rocks.

While his students listen, close their eyes and listen, squint in puzzlement and listen, Hopp is quick, identifies call and song as well as a connoisseur finesses his grapes and vintages. And like the adventurous taster he's always on the lookout for a discovery, something odd, rare, or distinctive. For instance, the huge zone-tailed hawk that soars in sweeping circles above us. A bird, Hopp tells us, that has found that by holding its wings up in a V shape it resembles a turkey vulture, a bird that the hawk's prey don't fear.

In the cool of the evening Hopp hangs a mist net—something like a hairnet the size of a badminton net—among the trees. He then steps into the shadows and begins playing a tape recording of his intended capture, an elf owl. In the dark the short call seems to have no direction. Repeated at steady intervals it seems less like several calls than one call that drifts off and returns. He stops the tape and we stand silently in the dark awaiting a response. None. He starts the tape again. But then between calls there comes a second call, as if the first call has returned too soon. We hear it. We hear it again. And then it stops. Hopp turns off the tape and shines a flashlight into the net. Ensnared in the fine netting is a gray brown owl the size of a sparrow with large golden yellow eyes. Hopp wraps a gentle hand around it and untangles the net from the bird's legs. Then he grasps the legs between his forefinger and middle finger so the bird seems to be perching on his hand. It is not struggling. This is one of the world's smallest birds of prey, North America's smallest owl.

Wintering farther south in Mexico, in the spring the elf owl nests and breeds in Arizona and migrates north to New Mexico, taking up summer residence in oaks or in saguaro cacti. Its range here is diminishing. Farming and housing developments like the ones that sprawl across the Green Valley push the bird into ever narrowing refugia. Once everyone has had a good look at the bird, Hopp releases it into the dark, where soon we hear it, or another, calling.

Glowing amber house- and streetlights from the retirement communities in the valley below look like small fires from thousands of encampments. Out here, way out here in the middle of the desert, the developments were required to use amber lights because their white lights were so bright the observatory telescopes couldn't make out the night sky.

The next day on our way out we stop at a desert wash, the sandy bed of a river that flows only when it floods. Hopp wants to see if he can find any of the Bell's vireos ("zheedle, zheedle, zhoo") that he's banded over the last years in these very thickets of mesquite. Small unassuming birds with drab gray-green plumage and white undersides that sometimes brighten pale yellow, they winter along the west coast of Mexico and come north to breed, some of them as far as the northern Great Plains. Hopp has been a devoted student of the birds, but rough habitat along small streams where the vireos breed and raise their young have been vanishing, lost to water diversions, drought, and cattle grazing. The birds, as a result, are far less common and the Least Bell's vireo, a subspecies, is endangered. The students search through the mesquite and Hopp, happily, finds an active nest.

Since I'm not going to be very good at finding vireos I walk through the desiccated bed of the dry wash. Looking at plants and river stones I come across a pile of what appear to be pebbles but turn out to be dried beans. Bending to pick them up I see more and can make out the tracks of whoever dropped them. I show them to Hopp, who doesn't seem surprised. They must have been spilled, he says, by one of the thousands of Mexicans who try hiking through the desert to make it into Arizona, human migrants unprepared for the desert, many of whom die of thirst, heat, and starvation each year along the way. I follow the trail of beans through the sands of the wash and soon its 10-foot-high banks narrow my world down to a hot, arid passage as good as sealed shut by the unrelenting sun, and I realize how only a few hundred yards of desert might seem vast and unendurable.

April 27
13 HOURS, 20 MINUTES OF DAYLIGHT

I would gladly linger in Tucson, enjoying the hospitality of the Malusas and another weekly paella prepared by Abuela, but the ocotillo is telling me that summer here is closing in. The temperature is predicted to hang at 90 for the rest of the week—perhaps higher, perhaps longer. While Jim and Sonya perform their end-of-spring ritual of repairing their rooftop evaporator—an early, still effective, and more energy efficient alternative to air-conditioning—I pack up to move on.

My next planned stop is Boulder, Colorado, where I'm to head up 12,000 feet into the Rockies with a team of scientists measuring the snowpack. In their last correspondence they suggested that I might come sooner than I'd planned: they'd had only 40 percent of the usual snowfall and it was melting quickly. Now I'm being pushed along by an early spring.

I have a week to get to Boulder and so Jim pores over the map with me to chart a route north different from the one I came in on. This is not a simple exercise since Jim peruses a map like a chess master at a chessboard. Where you or I might see a representation of borders, roads, and cities, Jim sees a multidimensional landscape full of decisions to be made, with each decision at once narrowing and expanding the possibilities. Jim charts my course with the idea of always being able to change directions while still moving ahead. After much chin-rubbing he suggests that I drive east from Tucson, up onto the high plains across to El Paso, then north through Oklahoma—suggesting possible detours, hikes, and trails all along the way—through Kansas and Nebraska, and then west to Boulder. Satisfied, he lets me fold up the map (although as I do I have the feeling that he sees whole ranges of mountains collapsing in on themselves) and after fond good-byes for everyone, I'm on my way.

I'm on his route for about 20 minutes before I change directions. Not to be contrary, but in my head is the nagging memory of a sign I'd seen on my way into Tucson from Phoenix. I turn back north. I have to have a look at Biosphere 2, the ghost town of the future.

In Oracle, Arizona, 30 miles north of Tucson, I pull off onto a winding drive and follow the signs until a shimmering complex appears, a vision of a past futurist's future, a development of glare-reflecting white buildings topped with geodesic domes and ziggurats of concrete, metal, and darkened glass. The scene reminds me of something from the illustrated books on space travel that captivated my childhood imagination, painterly representations of the planets, passenger rockets, and moon bases whose unearthly and daring conceptions I would pore over each night in my bedroom and that, along with the words "weightlessness," "other worlds," and "future," held me spellbound.

The structures at Oracle provide an impression no less provocative, sprawling across the calf tan sands of the Sonoran Desert before a backdrop of eroded foothills and distant high mountains. The whole thing appears to have dropped out of the sky. It's anime art, a lost world that never existed, a glaring anachronism. Imagine opening a tomb and finding a mummy in chrome-framed sunglasses.

The place appears deserted. Not just deserted, but as if everyone had been vaporized. One expects ray guns and Andre Courreges's designer space outfits, but the woman who materializes at the front desk seems unarmed and is clearly no moon girl. "The tour," she utters in a low voice, "begins in twenty minutes." I nod. "Out there," she adds, pointing toward a shaded patio.

I buy a ticket and walk through a series of exhibit kiosks and a little souvenir shop. No other visitors appear. I enter one of the vacant darkened little theatres, sit on a bench, and watch the history of Biosphere told in stills and film clips edited with a newsreel's urgent pace, earnest voice, and portentous sound track. This will be, I'm told, the story of a heroic scientific endeavor, "built ahead of its time," an "ultimate engineering triumph" to replicate Earth's major ecosystems—seas, coral reefs, marshes, deserts, rain forests—within a sealed and self-sustaining system where all air and water will be recycled. The plants and animals inside will live their natural lives and the humans will thrive on a small, but sustainable harvest. A spaceship Earth, this enclosed ecosystem will be the model for future space colonies and teach those of us left behind how to live a more sustainable existence.

Some rich men build model ships in bottles. Texas billionaire Edward Bass wanted to build a model Earth in a bottle. More, he wanted the model to be alive—with plants, animals, and, most of all, humans.

Outside, in the blinding afternoon desert sun, I meet up with the rest of those on the tour: a family of three and a grandmother with her teenage granddaughter. We're greeted by our stout guide Bill, who, despite there being only six of us, chooses to use his electronic megaphone.

"This is Biosphere 2. Biosphere 1," he informs us, "is the Earth."

As we approach the main glass and steel building, Bill itemizes the pertinent details: Completed in 1991 at a cost of some $150 million, the enclosed and sealed Biosphere 2 covers some three and a half acres. Six thousand, five hundred panes of glass cover the domes, which, when the system was up and running, held 7 million cubic feet of air, 30 thousand tons of soil, and 3,800 species of plants and animals. Above the 79-million-gallon aquarium, mechanical waves washed across a coral reef three times a minute. In the enclosed rain forest, sprinklers sprayed 120 inches of rain a year. There was a virtual coastal mangrove and a virtual coastal desert. There was even a quarter called the farm where the Biosphereans, as those who lived within the Biosphere chose to call themselves, could raise crops, pigs, and fish.

"It was," says Bill with ardor, "a twentieth-century ark."

Left high and dry in the 21st century.

"We had ten species of ants and five species of roaches. We had katydids, butterflies, moths, snakes, toads, lizards, and bush babies. They're all gone. Except the ants." Except, in fact, a single species of ant that killed off all the others and now, as we can see from the columns of them swarming everywhere we look, have the run of the place. I'd call them Biosphereants.

We follow Bill along a catwalk suspended 30 feet above the ruins of the ecotones below. Here and there a few hardy specimens persist—a banana tree that still bears fruit, a thorn acacia, a wiry frankincense tree, a mescal plant, and a hanging garden in leguminous tatters. He points to some barren ground where, he tells us, cottonwoods used to grow, sometimes as much as eight inches in a week. I have the wilting feeling I get when I return to my apartment after a long trip and see my dried-out houseplants.

He leads us down under the ecodome where large tanks recycled water from each of the ecosystems. Every tank had a backup tank. We stand at the edge of the quarter-million-gallon water reservoir designed to recycle the entire system once every two years. And then the true marvel of engineering, the great lung of the Biosphere—a black rubbery membrane the size of a circus tent. The flexible membrane allowed a 16-ton steel disk 90 feet in diameter to rise and fall with the daily increases and decreases of oxygen in the Biosphere.

The Earth itself does "breathe" like this. One of ecology's great discoveries was that during the spring, carbon dioxide levels in the atmosphere decrease because leafing plants use it for photosynthesis. Conversely, in the winter, CO_2 levels increase. The same thing happens, although not as dramatically, each day and night. As a consequence, in the closed system the Biosphere engineers had created, oxygen levels rose during the day as a result of all the plant respiration. And oxygen levels fell each night. They thought they had thought of everything—their creation could breathe.

In September 1991, eight Biosphereans, four men and four women, entered Biosphere 2 and the door—referred to, I guess, as an airlock—shut behind them. In July 1992, the Biosphere 2 Scientific Advisory Committee issued a "Report to the Chairman of Space Biospheres Ventures." Committee chairman Dr. Thomas Lovejoy (then assistant secretary of environmental affairs for the Smithsonian Institution and later to become chief biodiversity adviser to the World Bank) reported, "The committee is in agreement that the conception and construction of Biosphere 2 were acts of vision and courage." In September 1992 the team doctor pronounced all the Biosphereans, despite having lost 16 percent of their pre-Biosphere body weight, in excellent health due to their "nutrient-dense" diet. In October 1992 four Biosphereans cast absentee ballots in national, state, and local elections from inside the Biosphere. In October 1993, the Biosphere 2 crew came back out into the world of Biosphere 1.

Although all seemed to have gone right, amid the media hype and self-congratulations questions arose suggesting that much had gone very wrong. The "sealed" system had, in fact, been opened and adjusted many times during the course of the two years. Oxygen had been lost through the concrete foundation. An especially overcast Arizona winter kept the plants from peak photosynthesis and so they failed to absorb enough CO_2 and it built up inside to the point that Biosphere 2 underwent its own little nuclear winter—the greenhouse that suffered its own greenhouse effect. CO_2 scrubbers were secretly installed and yet oxygen levels continued to drop. For a time the Biosphereans lived with oxygen levels equivalent to those in the Himalayas. Food crops failed. Food intake suffered.

Reports came out that John Allen, the enigmatic Biospherean who first sold the idea to Bass, was more cultist than ecologist, more Barnum than Odum. A mining engineer with a business degree from Harvard, Allen was also a poet and playwright who wrote under the nom de plume Johnny Dolphin. The names of the entities Allen created to oversee Biosphere—Space Biospheres Ventures, The Institute of Ecotechnics—had the ring of science fiction. (Although Allen wouldn't be the last to take advantage of the instant significance "eco" or "bio" lends to almost anything.) To many scientists Biosphere appeared to be more theatre than science. Although for the few good years of its run the public, and not a few scientists, thought the Biosphere at Oracle might give them answers to some big questions.

Perhaps it was just something in the air. A few years ago geologists discovered that in Delphi, Greece, at the site of the 4th-century B.C. Delphic Oracle, faults in the rock of the mountain release fumes from beneath the earth's surface. Among these *pneuma,* as the Greeks called them, the geologists found ethylene, a gas that when inhaled causes reactions from euphoria to delirium to violent frenzy. So it turns out that the divining priestesses were probably high as glue-sniffers. The Oracle's prophesies were a ritual that required from those who sought advice a willing suspension of disbelief. Biosphere required the same.

A short time after it closed its doors, Biosphere contracted with Columbia University to take over the facility as an experimental laboratory. But after a few years the expense of running the place became too great and they quit the desert, concluding that Biosphere 1 was in greater need of their attention.

Another guide, Rod, a broad-chested man with a soft white mustache, tells me he was head of safety for the Biosphere.

"I once had a hundred people under me."

And now?

"Now there's only me."

Safety marshal and tour guide of a defunct biosphere.

As I leave a sour summery rain begins to fall. And as I drive west toward Texas across what may be the flattest land in the universe, I consider again Jim's basic approach to the world—"I always ask myself why things are where they are." Not why things are the way they are. Why not? Nature never has a time when things are what they are. In nature, unlike in the course of human events, beginnings and endings don't exist. Despite its appeal to ecology, Biosphere 2 was essentially a political happening, and in many ways a perverse one. To even conceive of the idea of shaping an ecological utopia, a bioengineered Eden, presumes some kind of actual control over—well, over everything. It was an attempt at ecological eugenics.

That this $150 million experiment could be wrecked by a

cloudy Arizona winter speaks of the dubiousness of intelligent ecosystem design and the hubris in thinking that we can replace a lost forest, wetland, or vernal pool. Even if you thought you knew everything that went into the making of a rain forest or a coral reef, the idea of capturing that place in time would be a fool's errand. Nature is neither a sealed system nor a scripted series. It has no intelligent design or designer. It's always open to the chance event, the singular moment: cell meets chloroplast, water drop meets rock.

For all of these reasons the scientists trying to understand what a warmer Earth, an Earth with an altered atmosphere, will be like know only that they can't know for sure no matter how many computer models they come up with. But not knowing with certainty is not the same as not knowing at all. Did the surgeon know for certain what the outcome would be had I not had heart surgery? He did know, however, that I had a damaged heart and that the damage was beginning to take a toll on the rest of my system. I thought this was knowledge enough. That humans have altered the Earth's atmosphere and that the altered atmosphere is beginning to take a toll on the rest of the Earth's systems is very clear, although often what's cause and what's effect, what's elemental and what's extraneous, what's primary and what's feedback, remain elusive.

For instance, should the melting of the Greenland ice add enough fresh water to the ocean so global ocean currents slow and the Gulf Stream no longer reaches north, the result could be a quick reversal of climatic fortunes. Oddly enough, after years of warming, the world may very suddenly turn terribly cold.

A report entitled "An Abrupt Climate Change Scenario and Its Implications for United States National Security" (written not by an environmental group but by consultants for the Pentagon) suggests imagining "the unthinkable."

There is substantial evidence to indicate that significant global warming will occur during the 21st century. Because changes have been gradual so far, and are projected to be similarly gradual in the future, the effects of global warming have the potential to be manageable for most nations. Recent research, however, suggests that there is a possibility that this gradual global warming could lead to a relatively abrupt slowing of the ocean's thermohaline conveyor, which could lead to harsher winter weather conditions, sharply reduced soil moisture, and more intense winds in certain regions that currently provide a significant fraction of the world's food production. With inadequate preparation, the result could be a significant drop in the human carrying capacity of the Earth's environment. The research suggests that once temperature rises above some threshold, adverse weather conditions could develop relatively abruptly, with persistent changes in the atmospheric circulation causing drops in some regions of 5–10 degrees Fahrenheit in a single decade. Paleoclimatic evidence suggest that altered climatic patterns could last for as much as a century, as they did when the ocean conveyor collapsed 8,200 years ago, or, at the extreme, could last as long as 1,000 years as they did during the Younger Dryas, which began about 12,700 years ago.

The cold will mostly affect North America. Some regions in the southern hemisphere may actually become warmer. There'll be water shortages, food shortages, more floods, and more droughts that "could potentially destabilize the geo-political environment, leading to skirmishes, battles, and even world war."

A real oracle would be useful right about now.

4

Geography Is Destiny

April 29
13 HOURS, 22 MINUTES OF DAYLIGHT

On the high mesas of the Chihuahuan Desert the soaptree yucca have just begun to bloom. These shrubs can live for 250 years and grow to 20 feet tall. Clumps of long slender leaves grow out the top of the trunk like an unruly head of hair. Come spring each clump of leaves sends up a thick stalk that produces crowded columns of cream white, bell-shaped flowers. Natives of the Southwest used the leaves for basket making. The flowers are edible and the large taproot was used for making soap. In the gathering dusk the tousle-headed silhouettes of hundreds of yucca begin to look like a carnival of Kokopellis, the dancing hump backed flute players of Southwest Indian lore. Some of these very plants may have seen just such ancient celebrations.

April in the Southwest is buzzard time, when the black and turkey vultures arrive from farther south and begin nesting and brooding by the thousands. A vulture, however, is not a buzzard. They were called buzzards by English settlers who'd never seen a vulture and figured it was a bird of prey like a hawk. In 1994 it was discovered that vultures aren't even vultures—at least what the Old World considers a vulture. DNA studies showed that vultures in the western hemisphere, even though they have pretty much the same repulsive eating habits as vultures abroad, are actually members of the stork family.

I drive northeast from El Paso. Ahead of me looms the long ridge of the Guadalupe Mountains with scored faces that seem to have been hacked at by a giant's chisel. Except for the increasingly unsparing light and heat I don't know how anyone could recognize a changing season on this barren, bone white expanse where the least gust of wind can kick up a spinning twister that will pirouette for a mile without running into anything. To the Aztecs such twisters were the work of Ehecatl-Quetzalcoatl, the bird-beaked god of wind in charge of sweeping the fields to attract the rains. A roadside black letter board, the kind that stands outside churches (although there is no church behind it), reads, "Pray for Rain. Thank You."

I can see El Capitán. What looks like the remains of a medieval castle is a lone raised ridge of reef limestone, 250 million years old. I've seen such formations in the Andaman Sea. Formed beneath the ancient sea as coral beds, over the course of millions of years lime-secreting organisms harden the coral into limestone. Once forced up out of the sea they crack, calve, and erode into haunting karstic presences full of fissures, hidden streams, and caverns. But the last time I saw one I was in a kayak, not a Chevy.

A historical marker comes up and, wondering what could have happened here worth a little roadside cul-de-sac, I pull over and read: "Resentment over control of salt lakes in this region often called Guadalupe Lakes led to the El Paso Salt War of 1877 which entailed the loss of many lives and much property."

I turn and look around.

"What property?"

There's nothing for miles. Who would fight over this? And where did they all go after the fighting ended?

As it turns out, the story begins here during the Permian epoch, when El Capitán is being formed, when a vast tropical ocean covers the entire region from Texas to New Mexico. Beneath lies a 400-mile-long coral reef (twice the length of today's Florida Keys' reef). With changing climate the ocean evaporates and mineral salts fill the basin and cover the reef. Over the next millions of years crushing geologic forces beneath the continent raise the reef, raise it thousands of feet into a limestone escarpment. As rains erode the limestone it runs down into ponds below. When the ponds evaporate they leave behind deposits of pure salt on bare white flatlands in the shadow of the eroded coral cordillera—the Guadalupe Mountains—at 8,700 feet the highest peaks in Texas.

Such deposits of salt were highly valued before the age of refrigeration. Caravans carried it from the desert lands of the ancient world and ships bound for Europe sailed with it in their holds. Here, however, only the Indians and the first settlers—most of whom are Mexican, because until 1848 this was all Mexico— harvest the salt and they do so freely. But seeing the possibilities of being able to sell the salt to the new pioneers heading west after the Civil War, in 1868 a group of six investors decide to lay claim to the salt lands. When the locals object the district court denies the claim. Out their stake, they begin feuding among themselves until one of the Salt Ring leaders—Williams—shoots and kills another—Clarke. In turn, two of the others—Fountain and French—shoot and kill Williams, appease the locals by bringing in on the deal a Mexican priest, Father Barrajo, and his friend, a local leader named Cardis, and then see to it that the district judge who ruled against them is replaced by Charles Howard, a carpetbagging Missouri lawyer.

In 1877 the salt flats seem all but secured until Cardis and Howard feud. Howard claims the land is his, blockades the road, and begins charging fees for the salt. Cardis objects. He confronts Howard in San Antonio and Howard beats him up. Not long after, he has another run-in with Howard, this time in Austin, and gets beat up again. At the salt flats, Howard arrests two Mexicans for taking salt without paying. Howard, arrested for imprisoning the Mexicans, is released on his promise to get out of town. He doesn't stay away long. He returns with a shotgun, finds Cardis sitting in a rocking chair on the porch of Solomon Schutz's store on San Francisco Street in El Paso, and shoots him there, point-blank.

With his friend Cardis dead, Father Barrajo puts together a posse to find Howard. Schutz telegraphs for help from the army stationed in Fort Bayard. The army sends troops. Howard, never without an angle, telegraphs the governor and tells him there are rumors of a Mexican invasion. The governor notifies the Texas Rangers. They all meet up OK Corral style at nearby St. Elizario. Howard and 20 Rangers find themselves outnumbered by Father Barrajo's mob. Word is that the priest has told them to finish off the Anglos. Howard and the Rangers retreat into a nearby house, to which the mob immediately lays siege. After two days Howard and the Rangers surrender. Barrajo's vigilantes execute Howard and two of his gang, mutilate the bodies, and throw them down a well. In a rampage afterward, St. Elizario is looted. The soldiers arrive. Howard's killers escape to Mexico.

Call it the Salt Wars allegory. The Earth's resources took billions of years to form yet by all accounts we, the world's peoples, are using them up at a rate that may leave us warring over what remains. Even under the best of all possible future climates, human needs for fresh water, arable land, and replenishable resources from the forests and the seas will soon outstrip what's available. A recent assessment of the world's ecosystems by the United Nations based on the research of 2,000 authors concludes that we've degraded some 60 percent of the Earth's ecosystems, lost at least 20 percent of the world's coral reefs, and damaged another 20 percent. Since 1960 the world's nations have quadrupled the amount of water impounded behind dams while doubling the amount of water withdrawn from the world's rivers. Reservoirs now hold three to six times more water than flows in all the world's rivers. Seventy percent of all the fresh water is used for agriculture, and most of that agriculture—since 1945 more land has been converted to cropland than in the 18th and 19th centuries combined—depends on synthetic nitrogen fertilizers that damage fresh water and the species that live in it.

"Humans are fundamentally, and to a significant extent irreversibly, changing the diversity of life of Earth," the report concludes. And while the changes now affect the poor most, the rich have no reason to feel out of harm's way. Global climate change, atmospheric ozone, the collapse of fisheries, and the loss of forests will affect us all, especially when economic instabilities result in catastrophe (after all, the insurance companies that end up paying for flood and storm damage aren't owned by Third World nations) or conflict.

Geography (and geology) being destiny, nations and corporations will lay claim to diminishing resources. (If you doubt this, know that corporations have already been set up to sell water when municipal supplies go dry.) The question is whether the world will deal with the problems or will Williams shoot Clarke, French shoot Williams, Howard shoot Cardis, and the mob kill the rest? I'm just thinking while driving. If the salt on these barren flatlands could incite a war, just think what kind of conflicts oil and water might engender. Somehow, I don't think six-shooters and shotguns will be quite sufficient.

The road north through the Texas Panhandle leads up onto the flat Llano Estacado, the "palisaded plain" that rises imperceptibly from 3,000 to 5,000 feet as I drive toward Lubbock. The palisades of the name is a continuous rock escarpment that runs along the north and west edges of the plain. The only feature that rises above the treeless flatlands, the ridge seems higher than its 300 feet and through the dry heat and haze has a miragelike presence, giving the impression that one might find water along its slopes—perhaps it's the reflected blue of the agaves. But aside from a hidden spring or two somewhere, there is no water here but for what lies deep beneath the ground.

May 1

May Day. Astronomically speaking it's midspring. In Europe, May Day marks the beginning of the summer season and as such has a long ritual lineage. As Frazer tells it: "It was and still is, in many parts of Europe, the custom to go out to the woods, cut down a tree and bring it into the village, where it is set up amid general rejoicings . . . The intention of these customs is to bring home to the village, and to each house, the blessings which the tree-spirit has in its power to bestow."

Frazer, of course, didn't live here, where there seem to be no trees between Lubbock and Amarillo, say, in the one-street town of Turkey, Texas, where it looks as if one of the Clanton gang might gallop in at any time. (Cowboy King of Swing Bob Wills was born here in 1915. Somehow he cottoned to jazz and the fiddle and came to lead his Texas Playboys to well-deserved fame and fortune.) In Turkey I get the feeling that the temperature reading is usually higher than the population, although today it's a wintry 40 degrees. This cold plateau is no place to celebrate the coming of May or to gain the tree spirit's blessings. Before the cattle grazed up the prairie and left it grassless as well as treeless the May Queen would at least have had the benefit of a wildflower wreath of bluestem and red and yellow Indian blanket.

Herds of bison once roamed these prairies in May. (Called *les beoufs,* they became known as buffalo.) Once they and the Indians—Kiowas, Comanches, whose livelihoods depended on the bison—were gone and the bones (of the bison) sold for making china or phosphate to decolor sugar, the cattlemen came. The biggest names in the business hereabouts were John Adair and Charles Goodnight, who by the 1880s owned 1,325,000 acres of land.

At Caprock Canyon, once part of the Adair-Goodnight spread, I come to the very edge of the escarpment. Beyond lies the rolling plains of Oklahoma. Below, however, lies what looks like a pit mine but is a brutal gulch carved out by 23 million years of erosion. I drive down, pack a sandwich and some water, and decide to take one of the canyon trails. Sometimes I'm hiking in sand, sometimes across broken rock. The walls of the canyon are layered with gypsum. This land, too, was once a sea.

The air is cold. The sun is hot. My only May Day fun is gazing at the spirited shapes of the hoodoos. I find myself thinking about the May woods.

Back east the forests have turned voluptuous shades of green. Migrating songbirds have begun to return, outglamorizing the steadfast but common titmice ("peter, peter") and chickadees ("chickadeedeedeedeedee"). Warbler song fills the woods. Scarlet tanagers ("chip-burr") posture on bare limbs. Juncos kiss the air and make mad dashes to and from their nests. Goldfinches ("perCHICoree") seem to hurl themselves in wavering flights through the trees.

This is the time, as one writer saw it in 1583, when "all the young men and maids, old men and wives, run gadding over night to the woods, groves, hills, and mountains, where they return, bringing with them birch and branches of trees, to deck their assemblies withall. . . . But the chieftest jewel they bring from thence is their May-pole, which they bring home with great veneration . . . covered all over with flowers and herbs . . ."

Hardly a stick of mesquite here among the desiccated hoodoos. I've been out in the desert too long. I don't even bother to follow a snake's trail through the sand or bend to examine a small prize of wildflowers, not when I know that back home among the old apple orchards I'd find the last flush of spring morels, light brown—apricot morels we call them. And that down in the hardwood wetlands where beaver dams have blocked the streams, grassy sweet young fiddleheads of ostrich ferns stand curled tight and ripe for picking. When the kids were young they used to go out into the woods and, along the banks of the beaver ponds, pick bushels of the fiddleheads that they'd bring back to Brooklyn and sell to one of the fruit-and-vegetable stands. They'd get five dollars a pound, and the vendor would sell them for twice that to those who had no idea where they'd come from, how they grew, and why they saw them only in spring— let alone that they'd been harvested by kids three, seven, and eleven years old.

Ah, the May woods, as Thomas Nashe wrote: "Then blooms each thing, then maids dance in a ring, / Cold doth not sting, the pretty birds do sing— / Cuckoo, jug-jug, pu-we, to-witta-woo!" Yes, I've been out in the heat too long. But the notion of May Day joy didn't translate to this side of the Atlantic. William Bradford, governor of Plymouth Colony from 1621 to 1656, had little good to say about May Day, especially deploring the Maypole revelries conducted by his neighboring colony of Merry Mount (on the site of what is now Quincy, Massachusetts).

"They also set up a May-pole," wrote Bradford, "drinking and dancing about it many days together, inviting the Indian women, for their consorts, dancing and frisking together (like so many fairies or furies, rather) and worse practices. As if they had anew revived and celebrated the feasts of ye Roman Goddess, Flora, or ye beastly practices of ye mad Bacchinalians."

The maddest of "ye mad Bacchinalians" in Bradford's view was the Merry Mount's governor, Thomas Morton, a hard-drinking Anglican, a lawyer who wrote ribald verse, America's original Merry Prankster who antagonized his neighboring governor whenever he could.

Drink and be merry, merry, merry boys;
Let all your delight be in the Hymen's joys;
Joy to Hymen, now the day is come,
About the merry Maypole take a room.
Make green garlands, bring bottles out
And fill sweet nectar freely about.
Uncover thy head and fear no harm,
For her good liquor to keep it warm.

Bradford lost patience. Despite his protestations that this new land was a place where one could worship freely, in May 1627, Plymouth Colony invaded Merry Mount, chopped down the Maypole, arrested Morton, and exiled him back to Europe. As Nathaniel Hawthorne imagined it some 200 years later, the "home of wild mirth made desolate amid the sad forest."

Perhaps Bradford would have felt more comfortable here in Texas rather than in a place where in May the berry bushes flower and the forest laurel and rhododendron will soon bring sprays of pink and white blossoms whose fragrance is winsome as first love. Too much Texas. I begin to trudge out of the canyon.

May Day has a miserable history here. In 1886, the Federation of Organized Trades and Labor Unions of the United States and Canada (later known as the American Federation of Labor) resolved that from May 1, "eight hours shall constitute a legal day's labor." They held a nationwide strike. On May 4, at a rally near Chicago's Haymarket, a bomb exploded, and what had been a peaceful rally turned violent. Police began firing; in the confusion, they mostly shot their own officers. Sixty were wounded and at least seven died. A citywide dragnet rounded up purported communists and anarchists, and four were eventually hanged.

May Day U.S.A.? Not likely. Especially after the Cold War. When I was growing up in the Eisenhower/Kennedy years all I knew of May Day were the black-and-white television images of the annual Soviet May Day parade: phalanxes of helmeted Soviet troops goose-stepping ahead of flatbed trucks carrying tanks, anti-aircraft guns, missiles the size of small spacecraft, and even the nose cone of a *Soyuz*. Martial music played and the drab parade passed before the tiered reviewing stand where the Soviet leadership, bundled and looking appropriately bearish in broad-shouldered alpaca, wool scarves, and fedoras (Moscow lies at about the same latitude as Labrador), surrounded the porcine premier Nikita Krushchev as if he were the goddamn president of the Groundhog Club's Inner Circle. The news on May 2 was given over to stern analysis. Who was standing where on the reviewing stand? Were they closer to Krushchev than the year before or farther from him? And what did where each man stood mean in the continuing Cold War? Any one of this inner circle—the Soviet, not the Punxsutawney one—might be the one "with his finger on the button" that we had all been told might someday be pressed, thereby launching the nuclear weapon that would force us under our desks at school or into the basement fallout shelter to live on soda crackers, powdered milk, and barreled water until we got the "all clear" to face the post-nuclear world—three to six weeks underground at most, we were assured.

May Day was Kafka gray. Hawthorne was right. "The moral gloom of the world overpowers all systematic gayety."

The Puritans among us remain alive and well, imposing arbitrary constraints on people or on nature (as if it cared) and seeking to control or extirpate the odd, original, unintended, or sensual. As it turns out, this canyon was a fine place for recollections of May Day.

May 3
13 HOURS, 41 MINUTES OF DAYLIGHT

Since leaving Jim's place in Tucson I spend my nights at a roadside motel or find a campsite and throw down my sleeping bag. I have a tent but the weather's been so dry that I can hardly recall what rain is like. Besides which, the car is so wide that if it does rain I can sleep on the bench seat. I'm not taking photographs because I've found it separates me from making a sensory connection with what I see. A good photo may evoke a great deal for the person who looks at it. It doesn't add anything to the experience of the person who takes it. Or at least if that person is me. Taking photographs of what I see when I travel would for me be like going to an art museum and taking photos of the paintings. Anyway, the spring world that I'm driving through is full of great art, lately all painted in sunlight.

In my cooler I carry the food I need for breakfast and lunch— peanut butter, jelly, crackers, nuts, some fruit and cheese—and then I usually stop somewhere for dinner. If it's early enough I'll pick up some meat, vegetables, or rice in a supermarket, and when I get to the campsite make a small fire if there's wood or cook on a little gas-powered tripod pack stove. One pot is fine. Boil water, use some of it for tea, then cook the rice, add the vegetables to steam through the rice. For frying I have a favorite cast iron pan, ancient and blackened—too heavy to take backpacking but I'm usually never too far from the car.

As I drive through Oklahoma, the herds of Angus cattle begin to look like steaks on the hoof and so outside Oklahoma City I ask around for steak house recommendations and I'm directed to a large restaurant in a strip mall. The tables are nearly full and the people for the most part blond and large. They have no tables for two, let alone one, so I'm seated at a table for four and given a menu and two slices of white bread and several pats of margarine in a red plastic basket. I just want steak and a baked potato but I have the right to a vegetable so I order okra.

At the table across from me a young boy of about 12 sits with his parents. They're all pretty big. The father takes a slice of white bread from the basket, tears it in half, pours on some ketchup, spreads it with his knife, and partakes. The son, watching carefully, then does the same. The waitress arrives with a plate stacked high with deep-fried battered onion rings that the father salts well before he and his son dig in. So far, the mother hasn't eaten anything. The onion rings finished off, father and son drink their iced tea from glasses the size of cocktail shakers. I drink my iced tea. It's mostly corn syrup.

My steak comes, a large but disappointingly thin cut of meat cooked too long. (One thing I've noticed in my travels is that outside the country's cities there's a fear of undercooked meat such that in many restaurants they refuse to serve anything medium-rare or rare.) The meat sits alone on the plate and the waitress has quickly gone. When she returns a minute later I realize why the order didn't come all at once. Each side dish fills an entire plate. One is piled high with enough deep-fried battered okra to feed a family of four; the other holds a baked potato the size of a paving stone. Already cut open, it's been dolloped with sour cream, non-pareilled with bacon bits, and confettied with yellow cheddar. Beneath it all it's hemorrhaging butter.

For me, food is a language. My grandparents spoke basic brisket. My wife, from the Eastern Shore of Maryland, spoke blue crab and Smithfield ham. As Heidegger said (or at least so my friend Miles told me Heidegger said), "Language speaks." So what I'm faced with here is not just a portentous meal but a problem in translation. What does ketchup on white bread actually mean? What is this dolloped potato saying?

I give it a try. We're a people who like things large—our side orders as well as our main courses. This is not news. We eat big, grow big, then climb up into our big cars. Of course we know that less would be better and healthier. We just don't know the real costs of the way we live. It may hit us hard when we find out, which may be when the rest of the world begins to live just as large. What amazes me is that even out here, close to the soil and grasses and grains from which the food comes, we don't appreciate food's earthy origins any more than do those who have the leisure to spend their lives dieting or those for whom food is just an inconveniently packaged nutritional supplement. What's the difference between thanking God for food or thanking the animal for giving itself to us for food? You don't dollop up something that itself has a soul, which one can address as a fellow being.

The waitress arrives. "Would you like to take the rest home with you?"

"No, thanks. I'm traveling."

Come spring Oklahoma is a hazardous meteorological intersection, a calamity waiting to happen. Winds blow in from every direction, bringing warm air, cold air, dry air, and hot air. Collisions are rare but when they occur—which may be a hundred times each spring—they can be sudden, fierce, and deadly.

Like anyone else I'd like to see such a ferocious natural event but be well out of the way when it happens. So I drive wide-eyed, imagining that I'm analyzing the cloud formations, and I listen to the radio for storm warnings, trying to remember if the right thing to do when the tornado strikes is to stay in the car—it's a big, heavy car—or get out and get under it.

On May 3, 1999, no one here had the luxury of considering their options. Even the Weather Service with its sophisticated storm-tracking models was caught unprepared for the pileup of storms and the explosion of tornadoes that followed. From 4:45 in the afternoon until 10 at night, 70 tornadoes dropped out of the sky. The worst of them grew nearly one mile wide and spun for 38 miles across four counties on a course that took it through Oklahoma City, tearing homes from their foundations, crushing cars, tossing livestock through the air, stripping the asphalt off the roads, the bark off trees, and injuring 295 people and killing 45. Winds of 318 miles per hour—the highest wind speed ever recorded—put the tornado only one mile per hour away from a ranking of six on the tornado scale that has never seen a storm above five.

When it comes to tornadoes per square mile, Oklahoma ranks higher than any place on earth. And they occur most often in May, when moist summery air from the Gulf of Mexico has moved north over the state. The seasonal transition might go smoothly were it not for dry air sweeping down from the Rocky Mountains to the west. It plows into the mass of Gulf air and pushes under it. Forced up high the moist air meets cold air. It expands, cools, and condenses. Clouds take shape. With more cooling they grow larger, wetter, and darker. The more moisture, the more energy. Inside, the huge smoky wall of clouds is full of violent upheavals. It is a massive moving mountain of energy.

High in the cloud, ice particles form that become electrically charged as they collide and fracture. Small positively charged particles hang up high. Large particles, with negative charges, fall. The energy ante rises. There's only air holding apart millions of volts of electricity. When the air's resistance gives out, Thor's hammer strikes the anvil. Lightning flashes. (Count the seconds until you hear the thunder of the shock wave, divide by five, and you'll know how many miles away the storm is.)

When the water droplets suspended in the air become too heavy, rain begins. If the updraft of air holds the moisture high enough for long enough chunks of hail form. The stronger the updraft, the bigger the hail. If the strong updraft begins to twist and contract, it can spiral down toward the ground—a tornado. This can take all day to come about. But because it's spring, the days are just long enough. Tornadoes inevitably happen in the late afternoon.

Storms such as the 1999 storm appear to come perhaps once every 20 years. The annual variations in the number and intensity of tornadoes is such that climatologists can't tell whether climate change has had any effect and enhanced detection technology skews attempts at making much of the historical record. On my way north through Tornado Alley, through North Texas, through Oklahoma, and into Kansas, I keep a lookout. Sometimes the horizon's foreshortened and threatening. But I don't even get rain. Despite the season's drought, irrigation keeps the low hills of northern Oklahoma so green that the state could be one huge fairway. Just put a tee down anywhere and swing away. The hills roll so endlessly that I begin to wonder if I'm driving on a treadmill. But when sunset comes and sets the tips of the grasses glowing yellow gold I can see that ahead of me stands a range of low mountains. The grasses grow higher and wilder. Buzzards enter the frame, moving north. A gleaming creek runs through a narrow wash crossed by a railroad trestle. Rising over it, in the lingering daylight, a full moon. I feel as if I've just passed through a gateway to the Great Plains.

May 6
14 HOURS, 1 MINUTE OF DAYLIGHT

The days remain cloudless. On the radio in Kansas they're talking about the blowing dust interfering with spring planting. Will China be buying wheat or soybean oil? The temperature reaches an unspringlike 90. The fields are too dry to plant. They're burned brown. All that blows through are "hopeful" clouds. Spring can be hard on farmers' hopes.

"We live by the numbers," a caller says to the radio host. "The price of fuel, the price of grain, and the amount of rainfall."

"But there are short-term numbers and there are long-term numbers," the host responds.

"This is the fifth year of this drought. To me, that's long-term and short-term."

Farmers here rotate crops of soybeans and corn. Bean farmers hope the United States will buy dry beans to ship to Iraq. Corn farmers hope that trade agreements will allow them to send corn syrup to Mexico.

On a far field a quarter mile away I see a rolling cloud of dust.

At the center of the burning haze I can just make out a man on a tractor, spring planting.

This has always been difficult land and the present state of drought nearly matches that of the 1930s.

> The nightmare is deepest during the storms. But on the occasional bright day and the usual gray day we cannot shake from it. We live with the dust, eat it, sleep with it, watch it strip us of possessions and the hope of possessions. It is becoming Real. The poetic uplift of spring fades into a phantom of the storied past. The nightmare is becoming life.
>
> A. D. CARLSON, *The New Republic,* 1935

The question is whether the nightmare will come again.

If I'm going to make it to the Arctic the last day of spring, I have planning to do. I can drive the Chevy only as far as Fairbanks—a long, long road—and since I intend to go up north of the Brooks Range into the trackless Arctic National Wildlife Refuge I'm going to need a plane. I check into connecting on commercial flights from Anchorage to Fairbanks and from Fairbanks to one of the communities on the edge of the refuge—Fort Yukon, Deadhorse, or Kaktovik—places whose names alone arouse the call of the wild. Once there, I'll need a smaller bush plane but I've got to reserve one now since the Alaskan spring is brief and the pilots get booked up quickly. June is prime time on the refuge: The threat of a late-season snowstorm is just about gone and the mosquitoes have yet to hit their maximum ferocity.

I make calls and find that a few informants recommend a husband-and-wife team of pilots, Terry and Terri Smith. Terry is a former Alaska Airlines pilot. I look them up on the Internet and I find myself as taken with their plane as with their credentials. I know nothing at all about aircraft but the Smiths fly what looks to me like a work of art, a Grumman Albatross, built in the early 1950s, that has the charisma of an industrial design classic, something from the drawing board of Rohde, Probber, or Loewy. Take a 1930s Zephyr locomotive and put wings on it. When I call them, Terri tells me that it is an amphibious plane, a "flying boat" designed for the navy in the 1940s. The Albatross can land on water, snow, and ice, as well as solid ground. There's none other like it in the world. And they'd love to take me up to the Arctic in it.

This all sounds great but as it turns out the Albatross is very expensive to operate. Even when we try to cut back on some of the refuge travel and Terri makes generous adjustments to their fees, the trip is still too precious an undertaking. She has another idea. Why don't I call Paul Claus? He and his wife, Donna, are good friends of theirs but friendship aside, she says that if she and Terry were planning this trip, Paul is the one they'd choose to fly with. In their estimation, Paul Claus is the best bush pilot in Alaska. And besides which, she says, knowing how disappointed I am not to be flying in the Albatross, Paul has a plane as distinctive as their own.

I find that hard to believe.

Paul's plane, she says, is also a one-of-a-kind—a DHC-3 De Havilland Otter, red and yellow with high wings and a single turbojet prop. Whatever that is. I thank her, give my mind's eye a few joyful moments with the Albatross before envisioning an Otter, and then call the Clauses.

To say that a bush pilot is one of Alaska's best is to say a lot since in Alaska bush pilots boast a status somewhere between gunslinger and Knight Templar. Anchorage has a museum dedicated to their exploits. Paul Claus, I find out, is a second-generation bush pilot (his father, at 85, still flies on occasion) and a shoo-in future Hall of Fame pick, right up there with Noel Wein, a one-eyed Fairbanks barnstormer of the 1920s, and Cordova, Alaska's, Merle K. Smith, whose singular misfortune during a particularly difficult takeoff earned him the sobriquet "Mudhole."

The last time I'd flown in Alaska I was embarking on a kayak trip through Glacier Bay. The plan was to fly from Gustavus, at the southern end of the bay, up to Muir Glacier and paddle back down—a journey of about a week. Once we'd packed the little seaplane with folding kayaks, camping gear, and food, there was not much room and a lot of weight. The two of us squeezed into the plane beside and behind the pilot and held our breaths as the plane bounced on the water a few times before it finally lifted off and headed up into the clouds where we couldn't see anything of the high peaks we knew stood all around us. My kayaking partner, a bit nervous but knowing better than to actually question the pilot outright, asked, "Where's the air traffic control around here?" The pilot lifted his forefinger and tapped his forehead. "Right here," he said. "Right here." We asked nothing more.

I'm eager to get back. The place had overwhelmed me with its big silent spaces. Its geology reset my sense of time: Sitting in a kayak at the base of a 1,000-foot-high glacial wall among drifting icebergs with their cobalt blue centers, hiking up and camping on the glacier, where the quiet is broken only by the whipcrack of the icy river's rigid current, and seeing bare rock that hasn't been exposed for 5,000 years, and the first green seedling to grow on that rock, was to experience the spring of a new world.

I had thought that Fairbanks might be far enough north to be the end of my journey. There, since 1906, on the night of the solstice the Alaska Goldpanners—"the farthest north baseball club on the face of the Earth"—have played their "high noon at midnight classic." But Fairbanks is still 160 miles south of the Arctic Circle and if I go that far, why not go to ANWR, the Arctic National Wildlife Refuge, which promised nothing at all: a place the size of the state of South Carolina with a population of exactly zero persons per square mile, some 30,000 square miles without a road, a car, a home or rental. There'll be snowmelt from 8,000-foot mountain glacier rivers high enough to raft, wildflowers in bloom, and birds that have made their own spring journeys this far (from as far away as South America and Africa), come to nest and brood. There'll be herds of caribou perambulating the vastnesses to bear and nurse their calves. The only thing missing will be sunset, but only because in the Arctic spring, due to the 23-degree tilt of Earth relative to the Sun, the Sun never sets and instead hangs at a midmorning resolve and makes a 24-hour circuit in the sky.

The Clauses and I work out an arrangement. I will fly into Anchorage, then take an air shuttle to Cordova (where the little airfield is named for "Mudhole" Smith). They'll pick me up there and before heading for the Arctic we'll fly to their lodges at the edge of the Wrangell–St. Elias National Park and Preserve. Happily, the Clauses had named their lodges "Ultima Thule."

May 8
14 HOURS, 8 MINUTES OF DAYLIGHT

I hear from the researchers in Boulder. Everything's set for the field trip up into the mountains.

"Make sure to bring snowshoes or cross-country skis."

I thought I'd packed just about everything into the Chevy. In fact, the Southwest and Plains have been so hot and dry that I've barely dug into the trunk for anything but clean T-shirts. Now heading north through Kansas for Kearney, Nebraska, before I drive west to Colorado, my good fortune is that Kearney, a city of some 28,000 that served as a fortified way station for pioneers on their own way west, has Cabela's, the Sears, Roebuck of hunting, fishing, and camping. Begun in 1961 as a kitchen-table mail order outfitter for fly fishermen, Cabela's is now listed on the New York Stock Exchange and sells shotguns, hunting bows, fishing rods, and all the gear and outfits to go with them. Underwear to outerwear, boots to hats, if it comes in shades of brown, green, or camouflage Cabela's has it. Even if you never hunted or fished in your life you can leave Cabela's looking like you have, feeling like you have, even leave smelling like you have, since they sell sprays that disguise the scent of human sweat so deer or bears won't be wise to your presence. Along with snowshoes, this is something for which there's no crying need in Brooklyn.

Despite it being spring, Cabela's store in Kearney has a pair of snowshoes on hand and on sale. Since it's nearly 90 degrees I don't wear them out of the store, but I'm prepared, at least so I think, for spring in the Rockies.

The snow I'll be shoeing over north of Boulder is the source of water for the South Platte River, which flows northeast across Colorado into Nebraska, where it meets up with the North Platte, which began up in the Wyoming Rockies and runs southwest. At North Platte, in western Nebraska, the two form the Platte River. A hundred miles later the Platte, now slowly braiding its way across the flat plains, wends by Kearney on its way to Omaha and its confluence with the Missouri. That is why I've come here: just as spring winds converge on Oklahoma so too do waterfowl on their northward spring migrations converge on the Platte River in Nebraska.

We are not speaking here of thousands of birds. We are speaking of millions. A half million sandhill cranes alone (90 percent of the species' entire population), along with millions of ducks, geese, piping plovers, and least terns. Most of these birds are on their way to Canada, Alaska, and even into Siberia to nest and breed. They've come from as far south as Mexico, some farther. Come early spring the birds begin to appear and from mid-February to the end of April the Platte is an overbooked avian caravansary.

Because the seasons change animals migrate. When the land or the sea no longer provides the food they need to survive, when the weather no longer provides the warmth they need to breed, animals must move or die. Nearly everywhere in the world people have marked the changes of seasons by these migrations. They depended on them for food. Farmers depended on the appearance of migrating birds to control insects while depending on the migrations of insects to pollinate their crops.

The time was when birds had seen much more of the world than humans; humans could only guess the birds' whereabouts before and after they appeared. Some thought the birds went into hibernation. Others assumed some kind of magic was involved. Migrating quail that pass through the Sinai Desert on their way from Africa to Europe most likely prompted the tale of heaven-sent manna that appeared to the starving Israelites on their 40-year journey. Barnacle geese were called that because in the Middle Ages it was thought that (perhaps just as frogs that hopped on land grew from swimming tadpoles) the great long-necked birds that seemed to appear from the sea began their lives as long-necked barnacles. This hypothesis caused not a little consternation in high places. After some priestly debate on the subject the Catholic Church allowed that since these geese were actually seafood they could be eaten on Friday. This put the rabbis of the time in a difficult position. They finally had to declare that Jews could not eat the geese because they were shellfish.

While migration seems like a good way to deal with seasonal change it does mean that, like having your kitchen in one home and your bedroom in another, survival depends on the upkeep of more than one place. If, for instance, while you're up in the Arctic nesting and breeding, the forest in Mexico where you spend the winter burns down, or the wetland where you need to rest and feed along the way has gone dry or been filled in and plowed up for a new farm or subdivision, you're out of luck. After 40 million years it is hard to begin changing your travel route. Many species have suffered from the loss of one or another of their necessary habitats. Monarch butterflies, for instance, which summer in the northeastern United States, have become threatened by the loss of the Mexican forests in which they winter. Songbirds and shorebirds that winter each spring in South America now return north in spring to the United States only to find that many of the forests and wetlands they need for rest, food, or breeding have disappeared.

Here on the Platte, spring once meant that snowmelt would

pour down the slopes of the Rockies, surge down the North and South Platte, and finally flood and scour the lands along the main course of the river. When the birds arrived they came to a place awash, studded with small islands. The turgid waters were full of seeds and nutrients washed downstream that in the increasing light and warmth brewed up a protein-rich invertebrate stew. It was spring: the birds ate, rested, courted mates, and could watch as a half million tall, long-legged, buff-plumed and red-crowned sandhill cranes began dancing.

Cranes dance. And when soaring, strikingly plumed, four-foot-tall birds that seem to come down to earth once a year appear and begin dancing, people notice. And since there are cranes that migrate across every continent except South America and Antarctica, many different people took notice and variously claimed that the crane brought them peace, happiness, justice, diligence, purity, loyalty, piety, filial gratitude, contemplation, wisdom, beauty, love, endurance, longevity, immortality, and marital bliss. And in places as disparate as Greece, Japan, China, Bhutan, Korea, Africa, Australia, and North America, people watched the cranes dance and danced like cranes. Or at least as close as they could come to dancing like cranes. A human might bow and bob like a crane, run, march with arms held akimbo and flap like a crane, even stamp, preen, and arch his back like a crane. But only a crane can combine all of these with 8-foot-high leaps into the air. Are they playing? Courting? Chicks dance. Adult mates dance. Courting pairs dance. And adding to the raucous convention, just before sunrise and at early afternoon feeding time, the garrulous birds start to croak, trumpet, bugle, rattle, and trill, alone or, most maddening of all, in unison. And they eat, digging into the river sands and waters for mouthfuls of microbiota, roots, and seeds. Over the few weeks on the Platte, sandhill cranes have to put on a tenth of their weight in order to make it up to the Arctic.

As we look out over the very dry May river flats, Paul Tebbel, manager of the National Audubon Society's Rowe Sanctuary outside Kearney, tells me that before 1940, before the dams, the Platte could provide more for the birds. Now there are just as many birds but their average body weight before heading out from the Platte has dropped each year.

"We have data for the last thirty years. They're in worse shape now than ever."

The dams hold the water in huge reservoirs and so most of the Platte River's late spring and summer flow is what seeps up from the ground. But what seeps up from the ground is not always good. Rainfall is negligible out here. And what does fall drains quickly through the sandy soils and eventually into the groundwater. How is it then, that even with the drought, this surrounding county had the nation's highest corn production? Irrigation spins water up from below. But because of the erodable nature of the soil a great deal of water is needed, as is a great deal of fertilizer. The nitrates and agrochemicals drain down and that contaminated water finds its way into wells and into the Platte. Without cheap water and government farm subsidies, however, there would be no farming.

"But farming," says Tebbel, "is all there is out here."

He appreciates the farmers' dilemma but he also knows that the Platte and the wetlands of the northern plains are critical for the birds' survival.

"We're the bottleneck of the flyway. Much of the rest is in bad shape. They can't go farther north without stopping here. So for the birds as well, this is all there is."

In order to appreciate more of the western Nebraskan landscape Tebbel recommends that on my way west I detour north through the sand hills.

The next morning I head out early, stop for breakfast in the little town of Ogallala, where a sign says it is National Prayer Day and where the bulky sheriff who struts into the restaurant wears his badge on his belt and a large wood crucifix around his neck. After coffee I head up to Route 2, which meanders through the largest sand dune formation in the Americas.

As the glaciers that covered the Rockies began to recede 10,000 years ago, winds blew the pulverized stone eastward, building dunes hundreds of feet high and miles long and sculpting them into a 19,000-square-mile temperate Sahara. Colonized and stabilized by prairie grasses, used for cattle grazing but useless for farming, the sand hills have few inhabitants. Ranches are miles apart but seem even more distant, hidden from one another by the swelling undulations of the dunescape. The road is a winding roller coaster ride, barely two full lanes, and full of blind curves and summits. When I pull off the main road my map is no longer any use. The roads seem to go where they can and for nearly an hour I don't see another car, home, or soul. This in itself is not unusual out on the Plains, but for some reason in the sand hills I feel like I've entered a maze and, perhaps because the road only twists, rises, and falls, for the first time in my trip I begin to question where I'm headed and why. I had sworn never to be lost but I'm lost, and because I can't see what lies ahead or behind, I feel a touch of panic, the kind of panic I used to feel when my heart went careening off rhythm and felt as if it were choking inside my chest. There was no pain, only a claustrophobia within that wanted out.

Even when I can get my bearings by the sun's direction there is no way along these roads to keep a heading. A road I choose because it leads me west will, in a mile, turn me east. The hills are dry, there is little green, and last year's parched stalks still cling to the fences into which they'd been blown. I decide to stop trying to navigate and instead follow the electric lines, figuring they'd have run them along the most direct route. There are no sounds out here. There is no radio station. I imagine for a few moments that I'm driving through a painted landscape. I wonder why this place has such an effect on me and run through the facts: There's no clear road to follow. I can't see what's up ahead. Feeling trapped I feel angry at every damn cutoff and dead end. Angry at myself for deciding to go off the main road just to see, *just to know.* Salvador Dalí used to say that he kept a sharp stone pressed into his gum because it felt so good when he took it out. That could be it. Does spring feel good only because winter is finally gone? I don't believe it. The more I know of it, the more of its life I comprehend, the more joy I feel at being a part of it all. I know how people could come to worship the Sun, recognize the light when we see it. Like the pilgrims of Merry Mount, like the cranes on the Platte, like lovers in May, I will dance.

The power line idea worked. I come back out onto the highway and head due west for Colorado and the Rockies.

5

Altitude-Addled

May 10

I thought I'd be free from the heat once I arrived in the Rockies, but I was wrong. Even as I drive west, upslope along the South Platte toward Denver's mile-high plateau, the temperature remains in the 80s. Mark Williams, one of the University of Colorado snow researchers I'm going to meet in Boulder, tells me when I call him that the Front Range of the Rockies experienced the driest March on record. As far as snowfall the winter wasn't great and spring so far has been a bust. I wonder if I can return the snowshoes. As I near Denver, exurban rangeland converges with subdivisions, intervals between exits off Interstate 76 grow shorter, and road signs require more reading. After weeks of meditative nosying around the open Plains I feel out of sync with the rhythms of six-lane traffic, as if I'm driving a farm tractor down the middle of the highway. Also, the midday sun shines through a nitrous haze that doesn't dissipate even as I drive northwest 30 miles to Boulder. The mountains are here somewhere but there is not a panorama in sight.

I get off one of those exits so crowded with motels that they share parking lots, check in, turn on the air-conditioning, get back into the still-hot car, and drive to the university campus. There I find Williams, burly and bearded, and his tall angular graduate assistant Tyler Erickson putting together the gear for tomorrow's trip to the Mountain Research Center. They work in a warren of offices and laboratories where researchers accrue acronyms like Russians do patronymics. Here one may work for INSTAAR (Institute of Arctic and Alpine Research) on NWTLTER (Niwot Ridge Long Term Environmental Research) or for NSIDC (National Snow and Ice Data Center), part of U of C's CIRES (Cooperative Institute for Environmental Sciences), which also includes researchers for NOAA, NASA, and the NSF. The director of the NSIDC also serves as vice-chair of the WCRP, which is the World Climate Research Programme, and member of GTOS, the Global Terrestrial Observing System. NSIDC also hosts the headquarters of the WDCG, the World Data Center for Glaciology. Boulder is the alpha and omega of cryospheric research, the study of sea ice and of snow, the amount of water and lands they cover, and their interactions with the atmosphere above them.

Most of us live in a narrow range of temperate climates. If snow and ice appear at all, we see them as relatively localized events. They are not. In winter, snow covers a third of Earth's surface. Ninety-eight percent of it falls in the northern hemisphere, where a quarter of all land remains frozen all year. Only 3 percent of Earth's water is fresh and more than two-thirds of the total fresh water is frozen in glaciers and polar ice. In their respective winters, sea ice anywhere from a meter to a few kilometers thick covers 14 to 16 million square kilometers of the Arctic and 17 to 20 million square kilometers of the Antarctic. This is the world of the cryosphere that the scientists study in Boulder. Because of the vastness of this world, what happens to it may be of far more consequence than what happens to the temperate world that we inhabit. This world is so inhospitable that research was nearly impossible before the 1970s, when satellites began sending back photos and telemetries from above Earth. What researchers have found is that the exchange of energy between cryosphere and atmosphere depends on some very fine variables, from the depth and chemistry of snow and ice to the contours and temperatures of the land and oceans that they cover.

We're going out tomorrow, explains Williams, to take depth measurements of Rocky Mountain snow to understand why its depth is what it is where it is and what that means to the eventual timing of the melting and the chemistry of the meltwater. My instructions are to meet back here at the NSIDC labs at 6:30 A.M. prepared to hike and snowshoe at 12,000 feet up along the Niwot Ridge.

That night, a warm summerlike evening, I pull my thermal clothing, snow pants, boots, gloves, windbreaker, sleeping bag, and snowshoes out of the trunk. I put some nuts and raisins in Ziplocs, fill the water bladders that fit into my pack, put binoculars, sunglasses, and notebooks in a belt pack. I've decided to skip dinner since I'd rather meet the altitude without a full stomach, and so instead pour a drink from the flask, close the cap, and drop it into my pack. I haven't done much but drive since leaving Arizona and I can't remember the last time I've been above a few thousand feet. I've been feeling fine and the doctor didn't seem concerned about the altitude before I left. I think to call again, but I decide it would be too much like stopping to ask directions when lost. Besides, what could go wrong? The heart valve works very much like the butterfly valve on a carburetor: blood is pumped through, then the flaps fold shut. Even the Chevy doesn't have a carburetor. Nobody has carburetors any longer. My guess is that once God had the mix of oxygen and fuel the way he wanted it he didn't have any problems with the occasional polluting emission. (He might have given it some more thought. Bacteria in the stomachs of cows and sheep produce methane as they break down food. Their burps, and to a lesser degree, their farts, make up some 20 percent of global methane emissions. And since methane traps 20 times more heat than CO_2, scientists concerned with global warming have been working on ways to cut back on ruminant eructation.) I'm just going to assume my valve job will hold up in thin air.

The next morning I meet several other scientists and assistants and we divide the group and gear among three cars. Erickson, a tireless quartermaster, checks his list of food and equipment and after a stop at his house to pick up a pair of ski poles I can use we head into the mountains to the university's research station, a state-of-the-art cabin in the woods at 9,000 feet with fully connected labs and offices. We meet up with the rest of the group, all but four of us under 30 and hale and hearty.

Since there's too much gear to carry, some of us, the over-30 few, will ride up with all the coolers, food, packs, and lab equipment in the only motorized vehicle that will make it up the mountain—an old hulk of a Sno-Cat with metal caterpillar treads that appears to have been rescued from a ski resort rummage sale. The young guys and one woman step into their cross-country skis, clamp their boots into their bindings, and head off on a gay glide up the mountain. The rest of us burrow in among the gear inside the jam-packed Sno-Cat and prepare to endure the hour-and-a-half trip up the mountain.

I'm in the back end inhaling exhaust fumes and being dislodged from my niche each time the machine lurches from snowbank to exposed rock.

"Sorry," the driver calls back. "There shouldn't be so much bare rock. But there is." Who knew global warming could be so hard on the spine?

I grab a ski pole to steady myself and take another and push open one of the roof windows to get some air. I find the tube from my water and drink in the hope of keeping hydrated and thereby forestalling the effects of the altitude. After a couple of rest stops during which I pee out most of what I've drunk, we reach the Quonset, a metal-walled hangar that's a combination weather lab and bivouac at 11,000 feet. The air is cold and clear, the sun glare is brutal and hot.

Once again we transfer the gear, this time from the Sno-Cat into the Quonset. Since it's already noon, Tyler opens some jars of peanut butter, some packages of cheese, bread, cookies, and we eat. With work to be done on the mountain Mark is anxious to get started. He hands his assistants long probes with which they're to measure the snowpack. Then he outfits us all with avalanche belts.

"Everyone wears one. It's spring and the snowpack can be unpredictable."

The belts hold electronic beacons that will signal our buried whereabouts if we become lost beneath the snow. We strap them on and test the signal. Everyone grabs his or her skis and heads out. I set my feet into the snowshoes, get up on top of the snow, and begin shuffling along behind. Our destination is the ridge at 12,000 feet and the glacier in the bowl just beyond.

The sunlight is bearing down overhead and the snow doesn't absorb a bit of it. I feel fine but as I trudge uphill I fall farther behind the graduate student crew, now energized with peanut butter and jelly. I'm moving slow enough to think—of my first days' gait after surgery, of my old doctor who'd told me that heading over to Europe on a troop ship in World War II he'd been so seasick that he was hoping he got torpedoed. I began wondering just how bad could an avalanche be. The doctor said the altitude shouldn't cause my heart any problem. He didn't say I might feel as if I were 20 years older. I try to ignore the sensation that the blood in my brain and extremities is beginning to congeal. My legs are leaden. I can't go on. I go on. And I make it up the last steep incline to just below the ridgeline, in sight of the glacier, where I sit down on the snow next to Mark. He'd already sent his damselfly crew winging off to take snow measurements, poking around the mountain's flanks as if it were some huge creature under examination, as if it were a living thing. As the students ski into the distance they become black specks on draped white canvas.

Mark and I both notice the incoming wind and the smoldering sky of an approaching storm.

Mark says, "Looks like you're going to get a taste of Colorado mountain spring." Since the students will be out working the slope for the next couple of hours I decide to get a head start on my way back to the Quonset in the hope that with some more slow walking my system will eventually find relief from the weight of the altitude. As it is now, I feel as if I'm towing the Chevy behind me. My assumption is that downhill will be easier going but with the patchy snow I find myself either having to scrape my snowshoes across exposed rock or take them off, hike, put them back on, strap my boots in, shuffle along. The uncovered tundra is brown as old straw and littered with dried blossoms caught midbloom in last year's first late-summer snowfall. Patches of turquoise and yellow lichens paint the rocks. Every dozen steps or so I need to stop. Not out of breath, but simply stultified.

I lean on my poles and look down the mountain to the foothill below and, in the distance, Boulder and beyond. Behind me stands the glacier peak, now enfolded in mist; the idea of it being a living being seems wishful thinking. It's too stolid and enduring. Nothing human about it.

I decide to go a dozen steps at a time until I come in sight of the Quonset. After 20 minutes I'm down to 10 steps at a time. It's nearly an hour before I finally see the base a couple of hundred yards downslope. I sit where I am, slide off my pack, and lie back on the tundra.

When the silence subsides I realize that I hear water trickling beneath the snow. I sit up to find out where it might be coming from and see beside me, nestled in among the rocks and glistening from meltwater, a tuft of wildflowers so small that it could fit in my palm. Being sessile and slightly delirious I decide that I ought to make a study of these 4 square inches of soil. In fact, I think, someone might base an entire life in science or art on this little spring efflorescence, where hair's-breadth stems support blossoms that couldn't bear the weight of a raindrop. I give myself until the sun goes down or until the storm that's brewing blows in. An hour ago it was blue above and glare below. Now sky and snow have merged into a depthless white. Gray, Mark had told me, will follow, and if gray, then smoky, then lightning that we'll feel first through our metal ski poles—energy flow, but not the kind you want up here, surrounded as we are by cold-crazed ions.

Just as I'd do if hunting morels in a spring forest I put my face close to the ground. This pocket-size refuge of soil, rock, and plants exhales musty terrariumlike warmth. If I can't hear the photosynthesis, at least I can feel the breath of absorbed light.

Snow reflects sunlight—which is why it's white. The researchers refer to surface reflectivity as "albedo." (Its Latin root, *albus,* means whiteness, and so gives us "albumin," egg white, and "album," a book of blank white pages.) The greater a surface's albedo, the greater its "whiteness," the more sunlight it reflects. Climate scientists spend a great deal of time measuring albedo because it's a key indicator of Earth's energy exchange. A perfect albedo is 1. Snow and ice can be anywhere from 0.6 to 0.9. A desert's albedo is 0.3, a forest's 0.15. The ocean has an albedo of only 0.1 to 0.6 and because there's so much more ocean than land, Earth's albedo averages out to 0.31.

Dark spots of ground like the one next to me have, say, half the albedo of the snow that surrounds them. The more sunlight they absorb, the more snow melts, and the more ground is bared. Gradually, the whole mountain's albedo drops. By next month the ratio of snow cover to ground cover will most likely be the reverse of what it is now. Plants (some just developing beneath the snow), will dominate the slope. They'll be absorbing energy, photosynthesizing, taking up some of the CO_2 that has been building up all winter. Which is all as it should be come spring.

But if the thaw begins too early and albedo drops too soon, things melt too fast. The effects may not be noticed in our temperate living zones, but the farther north one travels (or in the southern hemisphere, the farther south) the loss of albedo can have, and already is having, large consequences on the cryosphere. Because of Earth's tilt, the polar spring and summer growing seasons are more intense and short-lived. Up in the Arctic, for instance, where by the end of spring days are 24 hours long, small changes in albedo loom large. Every spot of exposed ground is absorbing sunlight for not 12 or 14 but 24 hours a day for nearly three months. In the polar oceans the effect is, as they say, the same but worse. When huge masses of ice break up, ocean water is left exposed. Since the ocean has a lower albedo than the land, the warming is quicker, the melting more intense. Worse, winds stir the heat into the ocean, and the stored heat continues to circulate even beneath the surface ice. So summer lasts longer, spring comes sooner, and the winter between doesn't last long enough to form its ice and snow covers. In the winter of 2004 the Arctic ice pack extended over 15 percent less of the ocean than it did 30 years ago. The diagnosis: low albedo. The prognosis: long-term changes in Earth's energy balance.

I don't know how long I've been malingering here exploring my miniature landscape but just as I begin wondering whether it was worth the trip to sit here like an old fool on the hill, there suddenly appear in my little terraqueous slough, from out of nowhere, like soloists from the wings, a ladybug and a black ant. In my altitude-addled imagination I decide that my little theatre of spring is complete. Animal life has arrived on the stage, and just as the stormy curtain begins to fall.

Back at the Quonset I lay my sleeping pad on the concrete floor to await the crew's return, but trying to lie down is hopeless. I don't feel tired, just restless and out of focus. I sit up, lean back against the wall, pull my knees up and look around. I can see that when there are no humans present, the hut is occupied by mountain rodents. They've left plenty of declarations of their intent to return. To my heightened senses the smell is intense. I stand and escape outside. If I'd brought my tent I'd gladly camp on the mountain.

Soon a couple of the young guys return, none the worse for wear. Then Mike Wireman, a hydrologist for the EPA and a veteran mountain guy, but one closer to my age, comes in actually looking like he's been out skiing high, steep terrain all day. He'd kept up "with the kids," he tells me, but now, he says, he's beat, can't stay overnight, and is going to head back down the mountain. Realizing that in my present state I'm neither going to be able to enjoy the good company nor endure the mouse shit, I tell him I'll go with him.

We fill up our water bottles, pull on our packs, and, with assurances from the young guys that it is only an hour and a half back down to the cabin, we head out, figuring we're certain to reach the research station before dark. We begin hiking and, since there is little snow on the first part of the trail down, Mike shoulders his skis and I lash my snowshoes onto my pack.

The air that has been bone-dry all day has begun to ice up. The wind is off the mountain and at our backs. Snow begins blowing in. When we hit deep snow on the trail Mike makes a ski run of it. I've got to strap on my snowshoes and trudge, sometimes through the forest underbrush, but Mike is gentleman enough to wait for me where the snowbank ends. Too bad I didn't ask if Cabela's had skis. When the trail again turns to bare rock we hike together. The path is easy to follow and once we're below the tree line the dense stands of ponderosa block the wind. We see lightning strikes ahead of us. Who knows how far? There is no sound but our footfalls on cold ground.

I find that just a few hundred feet of gravity make all the difference in the world. My legs move again the way I'm used to them moving and I begin enjoying my hike through the light, swirling snow. I haven't seen Mike for half an hour or so, ever since he took off again on snow that had drifted into the woods. I assume he's finding good surface and will keep skiing. I don't mind hiking alone. Somehow, when I'm surrounded by forest I always feel secure. Leathery pinecones festoon the trees; snowflakes fall lightly in the needles, glisten for a second, and then liquefy. Farther downslope the snow changes to sleet but I'm pretty waterproof and moving with more confidence all the time. I feel at home, actually, reminded of winter hikes in Pennsylvania.

I'm always sorry I don't get out into the winter woods more often. I recall once after a night of camping in the snow in the mountains of western Pennsylvania I reached the road to see if I could hitch a ride back to my car and was picked up by three 20-somethings in a 4 x 4, all of whom carried beer cans and shotguns.

"What were you doing out here?" one asks.

"Camping."

"In the woods? In the snow?"

"It wasn't snowing when I went out."

"Did you see any turkeys?"

"I saw tracks."

"That's what we're doing, hunting turkeys. We figure we can cover more ground in the car."

At least we had something in common. None of us were hunters.

I've been hiking for two hours and there is no sign of the base station. It has been more than an hour since I've run into Mike. I haven't seen enough snow either on the trail or off to ski or snowshoe for at least half a mile. The sleet has turned to rain and it's growing dark quickly. The question now is whether I wait for Mike or keep going. He may be waiting for me farther downslope. I try to recall if he had a full pack or a daypack. Does he have a headlamp? The trail's pretty clear even in the dark. And he is far more experienced in the backcountry than I am. But he was exhausted and skiing through the woods. He could have fallen. No way to begin looking now. How much farther could the damn lodge be? I'll keep walking. If he's not there when I arrive I'll figure it out then.

With darkness the rain turns back to sleet and then to icy snow. The wind now blows up into my face. I figure this means that I'm getting closer to the ridge where the lodge sits. I'm too tired to know why exactly I figure this, but it keeps me going. In another 20 minutes I see the clearing and finally, the lodge. It is locked up tight and Mike is nowhere to be seen. Since I drove up with the crew, if Mike doesn't show up, I have no way to get back to Boulder. If worse comes to worst I figure I can lay my bag out on the deck, and since the lodge roof has a few feet of overhang I'll at least be out of the snow. For the moment I sit on the porch, take some nuts out of my backpack, and drink the rest of my water. I figure I'll give Mike half an hour before I hike down the road to where I see some lights. But then he appears, a shadow on the trail.

"It was taking me so long to take off and put on my skis that I finally just decided to hike. But these ski boots are not made for hiking."

Twenty minutes later Mike and I sit in a woodshed of a nearby mountain bar too tired to finish our pizza or our beers, too tired to make even perfunctory exchanges with the very pleasant but hopped-up hostess who has told us, seemingly without taking a breath, that she owns the place with her husband who lived in Michigan but whom she met when she was in the merchant marine on a tour in Alaska where they lived for nine months, the good dark months, each year and this is their second restaurant, the first having been in Seattle, which is the home of the micro-breweries, but now that they're everywhere, like coffeehouses, they decided to live here and open up this place where the folks are all friendly. Mike and I just grin.

Two of the friendly folks, young long-haired dudes, stand at the bar. One, a White kid, has his hair in dreds. I hear him talking to his mate when the hostess goes to get our check.

"You know, it's like an hour to work in the morning and an hour back and work is like ten hours so that leaves me only twelve hours in the day for sleep and recreation. I don't think that's enough."

"But you got to go to work."

"I don't have to drive two hours."

"Right."

"I need my recreation."

Spring fever.

Mike and I get back in his car for the long ride down the winding mountain road back to Boulder. He is a fractured-rock hydrologist and the rock of the Rockies is very fractured. To keep himself awake as much as to enlighten me, Mike traces for me the patterns of the water flow through these mountains. But he knows them so well I think he could recite them in his sleep so I keep checking to make sure his eyes are open.

"The water flows through fissures in the granite. If there was a way to predict how much will flow and where it will end up—"

My head falls back onto the headrest. Suburban sprawl surrounding Denver has moved up into the mountains and, like demanding nomads, everyone wants water out of the rock. Adding to the problem is that contamination from 19th- and 20th-century mining operations still seeps through the fractures and into the water. I don't understand much more than this because I know nothing about water's subterranean proclivities or the geology around which it flows. Mike has got a whole other map of the world in his head, which is an extraordinary intelligence to possess. If only for a day I'd like to be able to observe what is going on above and below the surface at the same time, to see life in cross section just as it looks in one of those desktop ant farms.

I tell him this.

He says, "I like the practical applications of it. Water is important for people."

Of course it is, but Mike is not being modest or mundane, he is in earnest. When he says, "Water is important for people," it is not like you or I saying it for he knows exactly how difficult it may soon be to find fresh water. For Mike "water" is a worldview. If he was to say, "Water flows," it would be like Heidegger saying, as Miles told me he said, "Language speaks." In each case there is an implied demand for some deeper understanding and appreciation of something as elemental as life itself. Or it is life itself.

Yet we piss it away while in many places in the world much of a person's (for the most part, a woman's) entire day is spent getting fresh water.

Mike says, "Eighty percent of the water out here goes to economically unsustainable, even unnecessary, agriculture. There's too many people who want it and there's only so much of it."

Here in the Rockies, when deep winter snows melt in spring while the ground is still too frozen to absorb the water, the runoff will pour across the land down all the Rocky Mountain creeks, out to the prairies and the Platte. This provides water for species downstream. Cottonwood trees, for instance, depend on spring flows to spread their seeds along rivers, where these trees provide some of the only habitat in otherwise barren land. If the ground warms too quickly in spring it absorbs the snowmelt and less water reaches downstream. Which is why spring snows, even deep spring snows, do little to fill the rivers. A lack of winter snow—and the last years have seen only some 80 percent of the usual snowfall—compounds the problem. Worse are winter rains that add nothing at all to the mountains' storage of water.

Tonight, up on Niwot Ridge, more than a foot of spring snow falls.

May 14
14 HOURS, 20 MINUTES OF DAYLIGHT

The day after the long descent I feel fine. Still, I decide to spend another day in Boulder to get the car's oil changed and take advantage of an Internet café to catch up on news and mail and plan my trip to the Great Salt Lake. Nearly everyone I see in Boulder looks youthful and fit. The town gives bicycles and runners the right of way and just outside the town it is not unusual to see people dangling on ropes from rock walls. Seeing all the various outdoor gear on display, from high-tech clothing to high-tech bicycles, skis, climbing gear, and behemoth 4 x 4s, all got up with the glister of faux titanium, I'm reminded that while the last 10 years have brought increased commercialization to the outdoor experience, the actual protections accorded the natural world have been in decline along with this country's miserable record in dealing with the conservation of resources and climate change. We're capable of being a low-carb-bread-and-circuses people, easily distracted from the tasks at hand. But we also still hang on to the notion that it is necessary to balance protection of the environment with the protection of industry, as if these were two equal moral imperatives. The fact is that if you take away from business, new business evolves. Take away from nature and evolution dead-ends. When one crop fails, another might still grow. If the field lies fallow or is forgotten, a grassland or forest will replace it. But as a friend of mine, Bernie Herman, who studies vanishing agricultural landscapes, put it, homes, shopping malls, and parking lots are the last crop the land will produce. Alter the landscape by drilling or mining and that landscape is gone.

Before I leave Boulder I finally catch up with Koy Chounlabout, whom I'm hoping to see in the Oregon Cascades. I don't think she expected to hear from me. We'd met in the fall of last year when I was on an assignment in Oregon to write about the "invasion" of the national forests by thousands of wild mushroom pickers. The mushrooms, a species called matsutake, were wild; the pickers, despite reports to the contrary, were not. Most of them were Southeast Asian immigrants or the children of immigrants—Thai, Lao, Cambodian, Hmong—who came to the United States following the Vietnam War and who found that they could make a living harvesting mushrooms in the Pacific Northwest.

In rural Oregon they were greeted with more than a little xenophobia. Koy, when she wasn't out picking with her husband, worked for the Forest Service as an interpreter, cultural liaison, and troubleshooter. Short, stalwart, and assertive, she was born in Thailand, the daughter of a Laotian father and Vietnamese mother. She speaks Thai, Vietnamese, Laotian, Cambodian, and a few dialects between.

Just a few months before I had been kayaking in Thailand and the Thai wife of one of the guides introduced me to the Thai term *sanuk,* which expresses the Thai joie de vivre that takes pleasure from family, religion, business, conversation, and sustains the spirit in times of turmoil or, in Bangkok, perilous U-turns across oncoming lanes of traffic. As I understood it, this sense of the copacetic derives from an understanding that the material and spiritual worlds, being aspects of a single existence, are subject to the same truths. Thais don't find it irreverent when, at one of Bangkok's hundreds of shrines, they make offerings in the hope of a good payoff in the weekly lottery or that men in business suits gather in the parks in the morning to practice tai chi, focusing on the union of yin and yang, the passive and active aspects of the Cosmos, before going to work making deals. Koy, a compact, squarely built woman in a light blue windbreaker, her dark hair pulled back from her wide, softly angular face, could be sly with her employers, self-possessed with reporters, alternately defiant and loving with her husband, tough with her competitors, and keenly alert in the forest. For me, she embodied *sanuk.*

The pickers' migrations, she explained, take them on a yearlong journey, "the circuit," they call it, that follows the seasonal fruiting of forest mushrooms. After the autumn matsutake season they head for the Oregon coast to pick black trumpet and hedgehog mushrooms. Then they travel to Washington state and to Northern California for chanterelles. In March they take a month's break and then in spring return to Oregon for early morels, and from Oregon move to Montana for June morels, remaining there to pick summer huckleberries. They are not pickers for hire. They are not pickers out of desperation. Anthropologists who have studied them refer to them as a "community of mobile harvesters."

I told Koy that I'd be back in the spring, but she was used to being the main character in newspaper and magazine stories that she never saw by reporters whom she never heard from again. Journalists do a great deal of moving on. I don't know how many times I've become close to people about whom I was writing, promised to send them copies of the story, promised to keep in touch, and then did neither. But I could never do as much moving on as Koy and her cohorts, who, with intelligence and rigor, make their livings as nomads in 21st-century America.

She says she's been ill and had van trouble, but expects that by May 20 she'll be up in the forest outside of Sisters, Oregon, where, she says, I'm welcome to come and camp with her, her husband, and their crew. Because the pickers are wary of strangers I consider this a very gracious invitation. The spring harvest depends upon the forest fires that burned the previous summer. If all goes well, the charred soils in tens of thousands of acres of gloomy blackened woodlands will be bursting with morels, the first sign of life's return. But Koy says she heard that the Oregon harvest has been slow getting under way. If things don't improve they'll pick up and head off to Montana. But if I get to Sisters by the 22nd I'm sure to find them there.

I'm eager to get back into the cool damp forests. But before I do I have another long stretch of desiccated land to cross.

I pack up once more and leave Boulder heading north. My plan is to take the most direct route over the Rockies, Colorado Route 34 across the Continental Divide. The divide runs north-south along the high peaks of the Rockies and extends from British Columbia all the way into Mexico. Water that flows down the Great Divide's eastern slopes runs to the Mississippi and the

Atlantic; water on the western slopes runs to the Pacific. These days, wherever the water is bound, there won't be enough of it.

Instead of going through the narrow mountain roads I backtrack toward Denver and then drive north in the direction of Loveland and Fort Collins, where species-specific sprawl dominates. Suburban habitat around Denver now takes up a million more acres than it did in 1960. Over the same period of time the population has increased from 865,000 to nearly 2.3 million. Denver's burgeoning is part of the Southwest's population bloom. Between the 1990 and 2000 censuses, the populations of Utah, Arizona, Nevada, and New Mexico grew by 30, 40, 66, and 20 percent, respectively. Although the combined total of some 13 million falls short of the 21 million that live in the New York City metropolitan area alone, the average annual precipitation on the East Coast is 40 to 50 inches a year while in those four western states the average is no more than 12 inches. Most of that falls as snow along the spine of the Rockies and flows into the Colorado River, which is called upon not only to flow its natural course southwest but is also forced to flow east in order to irrigate the farmlands along the Platte. This last trick was accomplished in the 1950s by what the Bureau of Reclamation, bold as snake oil salesmen, called "a clever and complicated system that can make water do about everything—even flow uphill."

My passage across the Rockies will take me along the path of this "clever and complicated system," the Colorado–Big Thompson project (known in the West as the C-BT), which, as engineering feats go, is more Rube Goldberg than John Roebling, who built the Brooklyn Bridge. The Chevy is right now downshifting to keep its mass in motion as I drive up the steep two-lane winding mountain road toward Estes Park.

Building the C-BT took nearly 25 years and, as Wendy Nelson Espeland writes in her book *The Struggle for Water,* myth had as much to do with the bureau's work as politics. At the end of the 19th century, William Ellsworth Smythe, a Nebraskan newspaper editor who had witnessed the effects of 1890s droughts on farming, became a crusader for Western irrigation. Irrigation, he wrote, "was not merely a matter of ditches and acres, but a philosophy, a religion, and a program of practical statesmanship rolled into one." He later resolved that "irrigation is a religious rite. Such a prayer for rain is intelligent, scientific, and worthy of man's divinity. And it is answered. To put knowledge in place of superstition is the first step which men take in entering into partnership with God."

A publicity still often used by the bureau pictured an Idaho couple standing amid barren sands holding a sign proclaiming, "Desert-Ranch. Have faith in God and US Reclamation."

If the C-BT was God's work, the Bureau as His earthly agents went at it with as much machinery, dynamite, and political muscle as they could muster. They dug a 13-mile tunnel right through the granite of the Rockies so that spring snowmelt, collected in reservoirs on the western slopes, could be pumped through to the east into the Big Thompson River, a tributary of the South Platte, where it could power hydroelectric dams and be held in reservoirs for irrigation.

All went well for the first 20 years. Farming on the eastern slopes flourished. But in the age of consolidation, family farms still could not make a go of it. Farmland speculators gained ground, literally. Since 1950 the number of farms in Colorado has declined by 40 percent. Fifty years after the first water was pumped through the mountains, 3 percent of Colorado's farms are responsible for 73 percent of the state's agricultural business. Are they making money? In 1950, some 3 percent of farmers' incomes came through government subsidies; in 2000 subsidies made up some 70 percent of farm income. And even with the C-BT, recent droughts have cost farmers on the eastern plains millions of dollars.

The dam as avatar of the divinity didn't end with the C-BT. On its natural southwestern course the Colorado is impounded either behind Glen Canyon Dam in Lake Powell or in Lake Mead behind Hoover Dam. Together these two bodies of water are supposed to provide Arizona, Nevada, and Southern California with most of their drinking water and irrigate a million acres of farmland in California's Imperial Valley and another half million acres in northern Mexico.

As Lake Mead began to fill in with silt the solution seemed simple: Build another dam. When Glen Canyon Dam was being planned everyone from scientists to the bearded, bulldog environmentalist, novelist, and essayist Edward Abbey protested that the dam would destroy the Edenic Colorado River canyons and bring nothing but further unsustainable development to the Southwest. There was a flood of scientific papers, geological and biological surveys, federal and state reports, books, essays, and newspaper columns. The dam myth held out and Glen Canyon was built, making a wreck of the natural life of the Colorado. Spring no longer brought 85,000 cubic feet of water per second storming down the river, carrying along with it a half million tons of red silt and sediment that created habitat for wildlife and fish along the riverbanks. The river ran out from the dam, cold and blue-green from the lake algae, bringing with it little of any good-for-anything.

"To grasp the nature of the crime," wrote Abbey in *Desert Solitaire,* "imagine the Taj Mahal or Chartres Cathedral buried in mud until only the spires remain visible."

As I drive across the Southwest this spring the news is not good: the great man-made lakes that guaranteed the West water are lower than they've been since they were when first filling up. In Lake Powell, nearly 60 percent of the water is gone, the lake down nearly a hundred feet—you'd have to row your boat a half mile across sand before you could get it afloat. Abbey's sacred canyon walls that vanished 30 years ago have begun to reappear. Lake Mead, since it is fed by releases from Lake Powell, is not much better off. It is down 80 feet. Federal, state, and local agencies come up with either desperate projections or just-as-desperate solutions. Like madmen pulling out their beards, Las Vegas residents have begun pulling up their lawns to reduce the use of sprinklers. But the West, with its increasing population and declining precipitation, may soon be well beyond such symbolic triage.

Scientists who study the spring growth rings of ancient trees know now that the rule in the Southwest over the last millennium has been drought. The problem is that the commission established to dole out Colorado River water based their generous allocations on the 30 wettest years in the last 1,000.

To supply the water demands of all the Colorado Basin states, the bureau allocated a total of 17.5 million acre-feet of water. This was very generous, but it was also delusional since the river's actual average annual flow ranges only between 13.5 million to 15.0 million acre-feet. In a recent report the Bureau of Reclamation admitted that it might have overshot its mark. "The Colorado River is not only fully allocated; it is over allocated."

What will the Bureau do about it?

"The question of how to allocate the shortage between actual flows and the 17.5 million acre-feet in allocations . . . has not yet been addressed."

Tougher yet to address is that snow- and rainfall in the Rockies have declined 20 percent over the last hundred years, and the Environmental Protection Agency predicts that based on the last century's temperature increase, by 2100 spring weather may be another 4 degrees higher. Less winter snow, more winter rain, less albedo, earlier thaws, less water. It is pretty much a mess altogether and not one that will be solved with longer and bigger pipes. Especially poignant are the losses for a generation of farmers whose families moved west with the promise of land and water. And for the Native Americans who ended up on the driest lands and received none of the water at all.

Going west across the mountain I should get a good look at the source of the molested Colorado, maybe even glimpse the time in spring when, as a little informational film by the Bureau of Reclamation reports somewhat pruriently, "runoff climaxes."

(Water gushes to a crescendo of violins. The hands of engineers manipulate dials, the pumping begins, and the unseen narrator intones: "The barrier between has been pierced." Whew.)

I'll have none of it today. I reach the top of the Rockies only to find that the Fall River Road that leads across the Great Divide is blocked by snow. Estes Park, the little resort town for people who like to sit by the fire and drink hot cocoa in June, is the end of the line. I've got no choice but to go back down the eastern side of the mountains and find a way around. I decide to go north into Wyoming and cross the Great Basin to Utah.

May 17
14 HOURS, 37 MINUTES OF DAYLIGHT

I'm about to take the left fork, the Hastings Cutoff, as it was called in 1846 when, after crossing the Continental Divide at the Green River in Wyoming, the Donner party famously decided to split with the rest of their pioneer contingent headed west for California. Lansford Hastings, author of *Lansford Hastings's Emigrant's Guide,* assured the Donners and Reeds that the route south across the high Wasatch Range and down through the basin of the Great Salt Lake would cut time off their journey. The problem was that Hastings himself, a promoter not a trail guide, had never actually made the trip. The Donner party had to hack out a trail for their wagons, slowing their progress enough so that they found themselves in the desert in summer and in November got caught in the Sierra Nevadas during one of the century's worst winters. In the end, 41 of the 87 in the Donner party didn't make it. Those who survived did so only by eating those who died. But the very next spring Brigham Young and the Mormons followed the Donner wagon trail to Great Salt Lake.

My first day's drive across Wyoming took me across waves of plains beneath overcast blue-gray skies with winds blowing in from the west. The Great Divide Basin had plenty of green, and streams that appeared to be running high. All of this will change quickly and, very much like vernal pools in Eastern forests, the Basin's springs and meltwater pools will dry up come summer. The amphibians here also take advantage of the evanescent waters. The Great Basin spadefoot toad, for instance, responds to the first spring rains. Males emerge from their burrows, head for the nearest meltwater, and begin calling for the females. The females arrive, lay their sticky egg masses, and mate with the males. Once the adults leave, the generational turnaround is—has to be—quick. The eggs hatch in two or three days and the tadpoles develop before the pools dry up. Finally, much as the groundhogs do, they spend their summers fattening up and in late fall, in response to the changing hours of daylight, repair to their burrows for a winter's sleep.

When I cross the Green River I see how the Donners could have been enticed into heading south. The green-tinged tablelands seem eminently surmountable, promising possible passages through narrow mountain basins. Or the pioneers may have been, as I am, enticed and baffled by sudden appearances of gulls, circling and calling as if the sea were near. For the birds that arrive in spring the Great Salt Lake is a sea, the largest body of water between the Pacific Ocean and the Great Lakes, and from here it lies only a hundred miles to the southeast.

May 18
14 HOURS, 33 MINUTES OF DAYLIGHT

Near my campsite on the shore of Willard Bay, a freshwater extension of the Great Salt Lake, I sit on the sand beach, eat, and watch a raft of white pelicans line up shoulder to shoulder on a gravel bar, their long heads cocked every which way, looking dim-bulbed and cross-eyed. They're most likely getting away from the island in the lake, where there are some 17,000 of them breeding, doing a bit of clowning before they head back.

Sunset seems to be silly-bird time. An eared grebe swimming by looking like a little loon suddenly does a fitful 180 as if something beneath the water had nibbled on its feet. Every shrub along the shore seems occupied; among them I see shuttle orange-headed blackbirds, orioles, and goldfinches. Warblers pose in leafy nooks but don't like it when I lift my binoculars to get a better look at them. Magpies strut into every picture like bibulous wedding guests. Damselflies cruise among the wild roses, and huge fish—walleye?—roll in the shallows.

With the sun just about to sink below the Wasatch's snow-capped ridges a flood of red pours down their faces. The pelicans arise and gracefully wing across the darkening lake. The twilight is cool. I walk back to my campsite to make some coffee and find another guest has pulled in, a large RV with curtains in the windows and enough antennae to serve as mission control. They've got their umbilicals connected to the water, waste, and electricity, but no one appears. I pull my little folding camp chair out of the trunk, light up a candle lantern, sit with my coffee and, as usual, pore over my maps.

The birds' attraction for the Great Salt Lake is clear: If you were flying north from Mexico this is the first water you would find, and on this route there is none other between here and Oregon. The lake was once much larger, being a remnant of a freshwater lake, Lake Bonneville, which some 30,000 years ago covered much of Utah. When the lake lost its outlet to the sea and the Holocene climate became warm and dry, Lake Bonneville began to evaporate. The remaining rivers continued to pour in sodium and chloride from the eroding surrounding mountains and the result was a formidable stretch of desert, the Bonneville Flats, to the west, and to the east this big briny puddle spread across a playa basin.

For birds on their migrations north in spring, the Great Salt Lake is an oasis among the scarred and desiccated washboard of western lands—north-south-running mountain ranges separated by deep, broad desert basins—that begin here with the Wasatch, stretch west to the Sierra Nevada, and north to the Oregon Cascades. In the drier and cooler times of Lake Bonneville, lakes existed in all of these basins. But while species that had adapted to the cool and wet climate might have found refuge up in the mountains, migrating birds that came through only in the spring and fall found their basin lakes and food supplies—insects, fish, plants—vanishing.

Think of it this way: Every year you take the same vacation drive. You fill the tank, feed the kids, and head off knowing you'll make it to the next highway rest stop just as the car needs gas and the kids need a burger. But when you arrive you find the place out of business, shut down. So you begin to explain about climate change . . .

The size and salinity of Great Salt Lake depends upon the flows of the four rivers that feed it. In the 19th century the lake covered 1,700 square miles. In the dry 1960s the lake covered only 950 square miles. Floods in the 1980s spread the lake disastrously over 3,300 square miles. Since then the lake has continued to shrink and has nearly returned to its 1960s levels. Because the rivers keep dumping a million or two tons of salt into the lake each year, the salinity is now about four times that of the ocean. Nothing much lives in it—bacteria and brine shrimp. But in the spring birds arrive—spectacular numbers of birds.

I've not heard a peep from my neighbors but for the slight electrical hum of their big rig. I say "they" only because on the ground outside the RV's door sit two canvas captain's chairs. About 9 o'clock I hear a click of mechanical gears, and an awning above the front door folds back. Then—a Magritte-like touch—a pair of porch lights are turned on, illuminating a circle around the empty chairs. I wait to see if they're going to put out the cat. Maybe they'll have a newspaper delivered in the morning. If they came outside now I'd share a nightcap with them.

May 19
14 HOURS, 35 MINUTES OF DAYLIGHT

In the morning—I wake up around 7—the chairs and the RV have gone and I've once again got the campsite to myself. It's a nice place. So far Utah has the best camping facilities I've seen. I can pump fresh water and the very luxurious latrine has a sink and a shower stall.

All cleaned up I drive northeast of the lake and turn onto a narrow blacktop that runs across the marshes to Bear River Migratory Bird Refuge. On both sides water gleams among tall grasses and the place looks no different than any other healthy expanse of marshland through which I've traveled. All this is pretty good, considering what this marshland has been through. Native Americans knew how to make use of the plants and animals of the marshlands but settlers from the East knew only how to live by farming on dry land. By 1920 they'd built enough dikes to drain all but 3,000 of the delta's 45,000 acres. Migrating birds crowded into the remaining wetlands and became—like any animal, including humans, living in crowded conditions—easy prey for disease. In 1910, an outbreak of avian botulism killed 2 million birds. In 1928, with the official establishment of the refuge, canals were built to redirect the water and open up 25,000 acres of freshwater marsh. In 1984, nearly all these acres were destroyed when rising waters from the Great Salt Lake crested over the dikes and flooded the refuge with salt water. Once again the dikes and canals were rebuilt, the salt water flushed out, and the marshes restored.

The refuge now spreads over 74,000 acres and there is only one road in and out. The intense light has the whole place hopping. While spring comes in measured time to the tundra, forest, or garden, spring on the marsh leaps in. The energy of flowing water augments that of the light. Nutrients and oxygen stream through, arouse the decomposers—worms, insects, bacteria—that turn the

remains of last year's grasses to slimy detritus. In doing so they create a fertile compost that allows the wetland plants to process
energy, produce and store food at rates as high as, or higher than,
any terrestrial plants. The food, in turn, brings the fish, muskrats,
insects, and birds. Millions of birds: spindly legged birds with broad
wings, high-gloss plumage, and long tapering bills; duck-billed
birds that paddle and bob, squawk and waddle; all come to partake
of the output. Some, like the avocets with recurved bills, come to
dance as well. In spring their breast feathers blush to a pinkish tan
and they perform a crouching, craning two-step on their long blue
legs. I see black-necked stilts and double-crested cormorants. I stop
by a little pool and stand over two coots circling and clucking,
ignoring me completely. I walk along one of the dikes and see
snowy egrets, white-faced ibis, and another shelf full of pelicans.
Surveys of Great Salt Lake have counted 40,000 avocets and 30,000
stilts there for breeding; in a single day they've counted 600,000
Wilson's phalarope and 300,000 red-necked phalarope. Breeding
birds, migrating birds, some remain to roost while others, com-
pelled by the light and the lateness of the season, will press on
north.

As will I, since I've got to meet Koy in the Oregon Cascades
in three days.

6

$\mathcal{M}$ountain $\mathcal{M}$an

May 21

14 HOURS, 55 MINUTES OF DAYLIGHT

The map tells me that this spot in southeastern Oregon where the Chevy's battery has died is called Stinkingwater Creek, not an uncommon name in the West, where mineral springs often emanate sulfurous odors. This particular Stinkingwater Creek runs through Stinkingwater Pass, which itself runs through Stinking-water Mountain. I assume this is not a desirable address for I don't see a house within binoculars' view and I realize it has been a long while since I've even seen another car on the road. I passed the last town about 20 miles ago and it appears that Burns, the next town, lies just as many miles up ahead.

This morning I left a motel in Twin Falls before dawn, drove halfway across southern Idaho, northwest across the Snake River Plain (the route the Donner party didn't take) until I cut off onto Route 20 west into Oregon. I pulled over to take a break, got out of the car, stretched my legs, and while embracing the silence and solitude of the purple-green sea of sagebrush that surrounded me I noticed small bright patches of yellow wildflowers among the scrub. I picked a few, took them back to the car, got out my wild-flower guide, and began trying to make an identification. Now usually I can identify mushrooms even while I'm driving along at 50 miles per hour—but small yellow wildflowers? The book, with its little inset line drawings, was no help at all. Poking through the petals I felt as if I were trying to defuse a tiny yellow bomb. My final conclusion was that these were, in fact, very lovely yellow flowers. The problem was that I came to that conclusion too late.

I'd left the car lights on, and the battery was dead.

The remoteness of the sagebrush sea has since taken on a whole different aspect.

The Chevy and I know very little about each other's internal workings. But in the hope that I might find something jiggleable I raise the white hood and peer into the Dick's giant maw. If any-one does pass by it will look like I'm doing dental work on a shark. The battery connections seem fine but I nudge them a bit with my thumb. I wait. And then I try the car again—nothing.

I think after heart surgery you can be either eternally fearful, which complicates life a great deal, or decide that you can't account for every possibility anyway so you keep things simple. I am certain that some pioneers on their way west faced worse cir-cumstances. I have water, peanut butter, and a sleeping bag, and I don't expect snow for a few months.

Besides, I like this scenery. A flat plain bulging with mountains that were terraced by the lapping of ancient waves and are capped with eroded rock crowns. This is the northern extent of the Great Basin, the western reach of what is known as the Sagebrush Sea, named for the woody purple-green species that covers these high desert lands. Sagebrush is a tough customer of a plant despite its being a member of the sunflower family. It ain't no daisy. Even in places like this high and dry desert, where summers are scorching and winters freezing, sagebrush never sheds its leaves and even has the guts to flower. It may do so grudgingly—they're tough, little, hardly noticeable flowers—but flower it does.

With the hood open for anyone who might drive by in the next several days to see, I take out a guidebook and read up on Harney County, Oregon.

Area: 10,200 square miles

Population: 7,609

I try to imagine what fewer than one person to every square mile of land means. Not only what it means to my chances of someone driving by with a pair of jumper cables, but also as a 21st-century demographic phenomenon. In order for Manhattan to have that same density, only 20 people could live there. In 1790, Manhattan already had nearly 1,500 people per square mile. Now it has 70,000 per square mile. And not that Harney County has escaped the kind of crime that comes from people getting in one another's faces. At one time, despite all the vastness of the place and the limitations of 19th-century weaponry, people got close enough to kill one another. Indians killed settlers. Settlers killed Indians. Indians killed settlers. The Indians were rounded up and exiled to reservations. Then cattlemen who decided they could take all the land they could fence in killed other cattlemen whose fences crossed theirs.

The weather is changing quickly and white tendriled clouds sweep in on cold winds. A light rain begins to fall. In the last hour two cars have passed. Neither stopped. Two women in a station wagon pull over when I wave but, while they have plenty of sympathy, they don't have jumper cables. The sea of sagebrush changes from purple-green to gray. A red pickup pulls over. The driver has cables and we start up the car. I thank him profusely, he pays homage to the Chevy as I shut the hood—"Great car"—and I'm back on my way.

This is the first night I've actually called ahead for a reservation. The Hotel Diamond has only a few rooms but, 60 miles south of Burns, it is the closest place to the Malheur Wildlife Refuge, where I expect to meet up again with the birds of spring. The only other place to stay is Frenchglen, a two-building town named for Pete French, a tough little nugget of a man with a big mustache who cultivated what he must have deemed his likeness to Wyatt Earp. French was 23 years old in 1872 when he came out to Harney County from California, driving 1,200 head of cattle, and within 10 years he'd fenced in tens of thousands of acres of the county's grazing land—much of it arguably in public domain and not his to fence in. This did not please his neighbors. The most displeased was Ed Oliver, a rancher with a relatively small stake but whose land French had completely surrounded. After 10 years of feuding, Oliver finally took French to court and won a road easement through French's land to his own. French waited for Oliver to finish the road and then immediately sued to have it removed.

The day after Christmas 1896, French, from his farmhouse, saw Oliver approaching on horseback and rode out to meet him. Only Oliver rode away.

In January 1897, the Harney County, Oregon, jury came to a unanimous decision: Ed Oliver shot Pete French in self-defense. French then died of natural causes—a bullet in the head. Harney County can be tough as sagebrush.

All the auto stores in Burns where I might get a new battery are closed so I just keep driving, hoping the battery will recharge. I don't mind. The winding narrow blacktop, Route 205, turns out to be one of the most magical of all the roads I've traveled. Perhaps it's just the relief of being back on the move, but I'm feeling exhilarated by driving across what looks to me like open sea or, better yet, it's as close as I ever felt on the ground to the sensation of being in the air. In the vaporous twilight, the rolling desert and ashen mountains appear more cloudlike than terrestrial. And then the road rises up to a place called Wright's Point, so steeply that I see only sky until the car straightens and I'm startled by the panorama of an ancient inland seabed spread out below me. What is left of the sea are potholes and wetlands fed by snowmelt from the crown of mountains that encircles the basin—the Crane Creek, Blue, Riddle, and the Steens, from which the Donner und

Blitzen River flows (German for "Thunder and Lightning" and named by Colonel George Currey, who crossed it in a thunderstorm while chasing Indians during the Snake War, another story altogether). "North of me, south of me, spring is in flood . . ." and the sea of desert becomes a sea of grass surrounded by low hills of meadow hay.

The drive takes me around the refuge onto a narrowing back road that winds across the fields and then abruptly ends. Thankfully, the end comes just where the Hotel Diamond stands, a two-story wooden-frame ranch house with a screened-in porch. I meet David Thompson, who runs the hotel with his wife, Shirley, and Gretchen Nichols, and who also raises cattle on the land surrounding the hotel, land farmed by his family for the last hundred years. I take an evening birding walk with a few of the other guests and spot some avocets and stilts, yellow-headed blackbirds along the corral fences, hear, then see, an oriole, and best of all spot a languorous white-faced ibis. Everyone says they plan to get out to the Malheur Refuge tomorrow. But first, there's a dinner of barbecued ribs, my first home-cooked meal since Abuela's Arizona paella, and I can't decide what impresses me more, the food or my appetite.

That night, walking the road in the dark—very dark and overcast—I think more of what it might be like to live out here for a time, to grow used to this panorama. What parts of the brain that must be disengaged living in the city would kick in? This is not so far-fetched a notion since researchers have shown that with extensive use, certain structures in the brain may grow. MRIs, for instance, have shown that the navigation centers in the brains of London taxi drivers are larger than those in non–taxi drivers. Musicians develop those structures engaged by music. What would it be like to awaken one's dormant cells and senses to a new map of the world? I have this spring to try and find out.

After a sound breakfast the next morning—I have to move on if I'm going to make my appointment with Koy in the Cascades—

David jump-starts the Chevy for me and tells me where to buy a new battery in town. I'd like to spend some time at the refuge on my way out but since I can't stop the car I drive around it, stop where I can, leave the car running, and make brief forays. Unlike the much-diked and channeled Bear River Refuge, Malheur is like a country club for birds. There are already young geese following their mothers across calm pools, and twosomes of angular herons walk the high grasses. At the refuge headquarters tanagers stand out against the cold gray skies. Inside there is a chart up for birders to post their scores. The only thing missing is the sound of cleated golf shoes on the walkways. On the way out I stop again at Wright's Point. From what I've read a lava flow poured down across the land some millions of years ago, the current hardened in its place, and when the surrounding land eroded, what was left was Wright's Point. I recommend the view as highly as any, even as highly as the view from the Grand Canyon. This is a different kind of vista. Looking over the Grand Canyon one sees the hard edges between time past and time present; from the panorama at Wright's Point you can't tell past and present apart.

May 23
15 HOURS, 11 MINUTES OF DAYLIGHT

When the pistol came out, my son Nate's eyes widened and I sat up straight in my chair. It had been there on the table all the time but covered by the newspapers. Trouslou turned to the yard and fired twice in the direction of a cat, which did a turn in midair and then shot off toward the woods. Calmly, Trouslou put the gun back on the table. "Next time, he gets one of these." And he pulled out a small cowbell. "I love those cats, but they know I don't want them bothering the birds." All the other cats, I noticed, were wearing bells.

"What if you had hit him?" Nate asked.

"If I'd wanted to hit him," Trouslou said, "I wouldn't have missed."

"Cool," said Nate.

I am recalling this scene as the Cascades come into view above the ribbon of black macadam. The last time I was here, a couple of springs ago, I had been assigned to cover the battle over water in the Klamath Basin, just to the south. Nate came along with me and of all the things we experienced—these 10,000-foot volcanic peaks, the tens of thousands of migrating spring birds, canoeing in a spring snow squall—this scene in Greg Trouslou's kitchen remains, he always tells me, his most vivid memory.

Trouslou was an action hero, the kind of guy who looms large in a teenager's imagination. He was a backcountry guide with huge hands, his waterman's weathered look and cowpoke's leathery demeanor merging in a thick gold mustache. He told us he'd built his own house and led his own life.

At the newspaper-covered kitchen table that morning with his big mastiff leaning against his leg, Trouslou talked of the possibilities of maintaining Klamath's (and his own) refuge from the world at large.

Like Bear River and Malheur, Klamath is another high-basin oasis of lakes, wetlands, and marshes and also part of the National Wildlife Refuge system developed in 1903 by Theodore Roosevelt. At that time farming (much of it brought about by government irrigation projects) had begun decimating these Western wetlands. When birds arrived in the spring-market hunters shot them. Estimates at the turn of the 19th century were that some 6 million birds—pelicans, cormorants, great and snowy egrets, herons, osprey, and other waterfowl—migrated through the Klamath. It remains the largest winter nesting ground for bald eagles in the continental United States. Without these refuges and the protection of the inland waters they afford, birds' spring migrations would be near impossible.

The problem here in the Cascades is the same as that in the

Rockies: warmer winters, lack of snowfall, diminishing runoff. Here, however, the circumstances have grown dire. If the Bureau of Reclamation allows farmers to irrigate their lands to plant in spring, the birds on the refuge and the salmon in the rivers will suffer. And the farmers know all too well that without water for irrigation their farms will not produce, their fields will turn to dust, and their land will be worthless.

For Nate, the dilemma—people's land and livelihoods versus the value of fish and wildlife—was stark.

"It's like the movie *The Matrix*," he said to Trouslou that morning. "What you see is not what's really going on."

"I'm very sympathetic toward the farmers and their plight," said the usually laconic Trouslou, who recognized that Nate was hanging on his every word. "For generations these farmers grew up thinking they were doing the right thing here. They sweated over the land to provide us with food. But then, people have a responsibility to the place where they live."

Nate obviously thought these were wise words and I recall purposely keeping myself out of the conversation. Seventeen-year-olds have their own Tao. If a parent is lucky enough there'll be a few fleeting moments when you and he focus on the same thing. But you will break the spell of the moment quickly if you deign to offer an opinion or lesson about it. (To say the Tao exists, as the Taoists say—or don't say—is to negate its existence.) So a sentient parent learns to savor the moments of concurrency with his children and respect them with silence.

Rumors have it that if agriculture in the Klamath Basin vanishes due to lack of water, a plan is in the works to develop ski resorts on the forested mountainsides, the resorts replacing jobs lost on the farms. It's a dismal thought for both the farmers and these lush fir and ponderosa pine mountains, which are the very model of a million-year-old volcanic range in its still-volatile youth.

"I've lived just long enough to see what goes on," Trouslou

said. "And once change starts it snowballs. The land will become real estate—more development needing more amenities."

This, he said pointedly and to Nate's silent approval, didn't jibe with his way of life.

He had been staring outside as he said this and his hand had been reaching beneath the newspaper.

"No," he'd said slowly, "I'm just trying to lose track of time."

It was then the shots rang out.

In the years since, Nate has told the story many times and he doesn't have to exaggerate to make it worth the telling. "This guy's hands," he begins. "They were huge."

I'm glad to see these mountains once again, especially the forests. After weeks of burnt umber, green is a homecoming. To me, this feels like spring, enlivening my spirit and that part of my brain, whatever part that is, as devoted to the forest as a taxi driver's is to a city's grid.

I reach Sisters, Oregon, at 8:30 with the Sun. The little town, named for the nearby twin mountain peaks called the Sisters, sits right on the edge of the Deschutes National Forest, where Koy, her crew, and probably a couple of thousand other mushroom pickers are camping. The Forest Service office is closed so I can't find out which campgrounds have been designated for use by the pickers, and so I decide to just find a campsite for myself and hunker down until morning.

The map posted at the office shows the campsite locations. The closest area is only a couple of miles away and since it looks as if it's about to rain I go for it. The forest road is dark but I find the spot, get my sleeping bag and tent out of the trunk, grab a book and my flask, put on my headlamp, find the trail, and have to walk only a hundred yards into the woods before I reach the first smooth ground and a ring of rock around a fire pit. The wind is picking up and with the sun down the temperature falls along with the rain. I set things up quickly (realizing I haven't had to use my tent since my journey began), get inside, into my sleeping bag, pour a capful of whiskey and settle in with my book. But I put it down unread. The sounds of wind and rain through the trees are just too fucking wonderful. I turn off the lamp and fall asleep, listening.

May 24

15 HOURS, 13 MINUTES OF DAYLIGHT

In the morning I drive back over to the Forest Service office to find out where the pickers are camping. The ranger in charge of the mushroom harvest, a young crew-cut guy with a pearly goatee and green eyes that have a disconcerting symmetry, tells me the camps are not far and that he hadn't yet seen Koy (everyone knows Koy).

"How's the picking been?" I ask.

"Pretty good. We've had a little rain every night, which seems to set off good flushes the next day. They've been picking gray morels, blondes, and some king bolete."

The "blondes" are the pale apricot-colored morels. The king bolete is what in Italy is called a cèpe, an excellent mushroom, but in the eastern United States found only in the late summer and early fall. No relation at all to morels, boletes have caps but on their undersides hold their spores in a spongy surface of tiny pores rather than in gills.

"How many pickers so far?"

"Oh, we've had some thirty-five hundred or so, but on the weekends we can have another thousand on top of that."

"Mostly Asian?"

"Asian and Hispanic. About twenty percent White. They're from everywhere: British Columbia, Texas, Mexico, California, Washington—"

"Getting in each other's way?"

"Not this year. Last summer we had the biggest fire in the

Pacific Northwest. Ninety-six thousand acres burned and we've opened forty-five thousand of that for picking."

"Where are the buyers' tents?"

"This year we gave the buyers a place to set up right in the forest and we've got a special forest products checkpoint."

This, I knew, the pickers wouldn't like. For immigrants, "checkpoint" has an unpleasant ring.

I thank him for his help.

"You going to be picking?"

"Uh, yeah, I guess I better buy a permit."

Two dollars a day, a hundred for the season April to May. I buy a daily permit and head out.

Ninety-six thousand acres is about seven times the size of Manhattan. There's no way I'll find Koy and so I drive off to explore the forest and, since I have my permit, maybe find some morels.

To be fair to the Forest Service, dealing with thousands of mushroom pickers, Asian, White, or Hispanic, has not been easy. For years the mushroom-picking business in the Cascades consisted of a few local shroomers who sold to restaurants or to distributors in Portland. Then the matsutake rush began. A white-cap mushroom with a dense, macadamia nut texture, the matsutake's most distinguishing characteristic is an odor that has hints of strawberry and basil (it has in common with both the aromatic oil methyl cinnamate) combined with a whiff of aging cheese. It's not to everyone's taste. Norwegian mycologists call it *Trichloroma nauseosum*.

The Japanese love it. They serve it with ceremony. But as blight killed off tens of thousands of acres of Japan's red pines, the matsutakes that grew in the trees' shadows disappeared. The harvest crashed. The mushroom's rarity raised both its allure and its market value. Asian buyers seeking new sources of supply found them in the Pacific Northwest, where matsutakes crop up each fall in the mountain forests from Canada to California.

In the early 1990s matsutake mushrooms in Tokyo sold for as

much as $1,200 a pound. Pickers in Oregon were being paid $600 a pound. A single mushroom could be worth $100. In 1992, an ounce of matsutake mushroom was worth as much as an ounce of gold. Mushroom fever struck. Pickers swarmed by the thousands into the small, close-knit farming and timber communities of the Cascades and into forestlands over which the locals held an emotional, if not actual, proprietary interest. "Xenophobic" might be too unforgiving a word to describe the locals' reactions, but it would be close.

Local landowners came to the Forest Service with trespassing complaints; they threatened vigilante action. Pickers said they'd been shot at or had dogs sicced on them. Rumors of shootings in the forest over mushroom territory or money gained ugly credence after the murder of a Cambodian matsutake picker in 1992. The next June a Cambodian morel picker was shot and killed. The national press picked up the stories and portrayed rural Oregon as a lawless backcountry.

The Forest Service was flummoxed. Their fields of expertise didn't encompass social work and cultural anthropology; they knew the ecology of the forest. Without mushrooms, the forests not only in the Pacific Northwest, but also around the world, would be far poorer places—if they could survive at all. These mountainsides of pines and firs that I'm driving through might justifiably be called fungal forests as well as coniferous forests for the trees and the fungus are codependent, symbionts intertwined at the trees' root tips to exchange sugars (produced by the tree through photosynthesis) for nitrogen (produced by the fungi with the help of the bacteria they host) as they break down organic matter on the forest floor. Mushrooms, themselves unable to photosynthesize, recycle the sun's energy through the soil and back into the trees.

The Forest Service studies demonstrated that the mushroom harvest was neither damaging the trees nor the mushroom crop. In fact, the research suggested that if the value of the mushrooms remained high the mushroom economy in the Northwest forests would become more valuable than the sagging and environmentally challenged logging economy. What grew belowground would be worth more than what grew above.

In the burned forest there is nothing growing above. From an 8,000-foot summit I get out of the car and overlook steep charred moonscapes strewn with the blackened hangs of once sturdy trees. When I walk out onto the barren hillside gray soot coats my boots and I can feel the surface heat seeping through. The dark soil absorbs so much heat that the temperature on the burned ground can be more than 150 degrees. I scan the ground and morels appear, one after another, gray as the ground in which they grow. Picking all day for weeks in a row in this heat with no shade, being coated with soot, is not for the casual forager.

For the new professionals the Forest Service promulgated rules and regulations that covered what areas could be picked, where pickers could camp, how big the mushrooms had to be, how to pick them without damaging the mycelial network beneath the ground that connects the mushrooms to the trees. The service translated the regulations into Spanish and a few Southeast Asian languages. Those in charge made trips to the Asian communities in Oregon, Washington, and Northern California from which the pickers came. Koy worked as liaison. By the mid-1990s things had grown calmer. Local stores even began to see opportunity in the pickers' arrival and stocked Asian foods and rice.

And then the market began to decline. At the end of the 1990s Japanese buyers found new sources of matsutakes in the pine forests of Korea, China, and Tibet, where pickers would take less for the mushrooms they harvested. In 2002 a bumper crop of Korean mushrooms caused the Pacific Northwest market to collapse. Prices fell to $2 and $3 a pound. Pickers quit picking. Many pulled up stakes and never returned, leaving a hard corps of harvesters such as Koy who expect tough times but who have no intention of forgoing the freedom of life on the circuit.

I don't have a pack on so I pick a dozen or so morels, walk back to the car, and head to the campsites. The harvesters aren't back yet. They usually pick until sundown and then bring their day's take to the buyers, who will sort the mushrooms by size and quality, weigh them, and pay in cash. Buyers' prices fluctuate during the course of the harvest, sometimes during the course of an evening. Pickers can wait out a low starting price but only for a few hours. Mushrooms are mostly water and as they dry they get lighter. Morels dry very quickly. In fact, I tossed my day's take on the shelf behind the Chevy's rear seat and with the sun shining through the rear windshield they dried in a couple of hours, giving the car's interior the seductive aroma of a smoky latakia.

Tents are pitched, clotheslines and tarpaulins rigged from tree limbs, and fire pits dug on either side of the dusty forest road. Many of the harvesters live out in the forest for weeks at a time. Fall and spring temperatures can rise during the day into the 80s and drop at night into the 40s or to freezing. Even summer nights can be cold. In the evenings after they eat, the harvesters will often play cards, sometimes for money. For how much money depends upon how good the harvest has been.

One of the buyers is in her tent setting up her table and scale. I ask her if she has seen Koy. She had, she said, and her group is camping at the very end of the road. I drive down but no one is there so I turn around and drive back to my campsite to pack up my gear. Since I can see what the place looks like in daylight I take a walk through the woods, find a stream and follow it for about a quarter of a mile to its source, a gushing of water from out of the ground hidden within a thicket of butter yellow flowers. I look at my map and see that this sweet little expression is the spring of the Melior River, which broadens into one of the Northwest's finest trout streams, running for 20 miles until its confluence with the Deschutes.

By this time I see the transfer of energy everywhere I look and as the season warms I begin to recognize dynamic systems taking form—light, spring, flower, river, mayfly, trout, angler—and I partake, not feeling my gradual separation from the world outside.

When I arrive back at the mushroom camp I find Koy's white van parked in the woods at the center of a semicircle of tents and pickup trucks. I pull the Chevy into an open spot and as I do, all the activity in the camp stops, waiting to see who this is. I get out and spot Koy, and everyone relaxes when they see that she greets me warmly.

"You drove all the way out here?"

"I told you I was driving. You like the car?"

"Must take a lot of gas," she says. "Are you staying here? With us?"

"If that's okay with you."

"Do you have a tent?"

"I do. I'll set it up over there."

"You know that photographer came to take pictures and he didn't even stay for a second. No tent. No sleeping bag."

"You know. Photographers just have their equipment."

"He wasn't a very nice guy."

Koy has a way of putting things very matter-of-factly without any harsh edge. As if his not being a nice guy was just how it was and it was probably more his problem than hers.

The men are building a campfire and she makes introductions. I don't get all the names—Ainoy, Lit, Sam, Charlie—but we shake hands, I'm offered a cigarette, take it, and begin to help gather the firewood.

Koy is cooking, tending a huge wok of steaming broth and noodles being heated on a propane stove. She pulls a few plastic Ziploc bags full of marinating chicken out of a cooler chest, closes the chest, puts a cutting board down on the top, and begins chopping chunks off the chicken.

"You like spicy, right?" she asks, remembering that the last time we had dinner, in the fall, she and I and another photographer, Antonin Kratochvil, spiced things up to the point that I thought Antonin was about to pass out.

"Spicy, right," I said.

"Like Antonin," she said, laughing.

"Not like Antonin."

"I liked Antonin. He's good?"

"Antonin," I agree, "is great. He told me to say hello."

With darkness the wind rises and the men have to turn their backs to the smoke blowing up from the fire. I'm offered a beer and I take it and watch as Koy pulls out small bags of spice from a box, adds them to the chicken, and the chicken to the soup. I ask what spices she is using but she says she knows only the Thai names for them.

From the next campsite I see a woman approaching with a lantern hanging from her belt and carrying a large glass jar full of a murky liquid. When she reaches us everyone greets her, laughing, and she takes out a shot glass, and then goes from one to another giving each a shot of what's in the jar. She watches each one drink like a mother would watch to make sure her child finishes up her medicine. They drink, eyes blink, lips purse, and then they smile for her.

"Try it," Koy tells me. "It's for strength in climbing mountains."

"What's in it?"

She shrugs. "Bark, herbs, mushrooms—and very strong liquor."

I'm offered a glassful, nod my acceptance, and drink.

"Good?" the woman asks.

A very earthy schnapps.

"Very good," I say, and chase it with beer.

Koy puts out a large bowl of rice. She rolls ground-pork balls and adds them to the soup. Everyone has a bowl of soup and then takes pieces of meat out of a marinade, cooks them on sticks over the fire, dips them in chili sauce, and eats them with fingerfuls of sticky rice. No utensils needed.

As we eat I talk with Sam, a man who's my own age, a Hmong who tells me that in 1976 he'd been working with the Americans in Vietnam for only a couple of months when we withdrew.

Knowing the North Vietnamese would kill him, he escaped into Thailand, where he lived until 1979, when he made his way to the States.

There's a pause. English is not easy for him and I know that if I spoke Cambodian he'd have a great deal to say. But he makes the choice of subject.

"We ought to go in and win the war in Iraq," he says. "You don't go to war unless to win. I am ready to fight. Older men know how to fight better than young men. Send old men. We know that when we fight we fight to win. Or we don't fight."

We drink. The wind shifts and we turn our backs to the fire.

Sam says, "We're okay as long as the wind is blowing. If the wind is blowing it won't rain."

"You like living out here?" I ask. And his answer came in broken phrases so I'm going to paraphrase here rather than make him sound like an illiterate child because he is not and he has lived out here and, like me, he is no longer young and so doesn't talk through his hat.

"It's become very hard lately," he says. "I have to pick a pound of mushrooms to pay for a gallon of gas. That's not good. But I love the seasons and the trees and the mountain air. Without nature, there is no human life." And when Sam says "human life," it's like Mike in Colorado when he says "water." I think Sam knows human life both above the surface and below.

After dinner a sheet of plywood is set across a couple of tree stumps and a card game begins, but it is not long before the wind picks up and the beer bottles placed on the cards to keep them from blowing away begin to shake as if there is a spirit arising beneath the table and then, to everyone's amusement, a gust of wind blows the whole table away, beers, cards, and all.

Other pickers from nearby camps have come over and Koy slides open the side of her white van to reveal, above her and her husband's makeshift bunk, a shelf of electronic gear that includes a 27-inch TV monitor, big speakers, and a couple of amps. Koy takes the microphone, her husband puts in a tape, and the karaoke bar in the forest opens for business. The singer on the screen sways to the music and words, in Thai or Cambodian, run across the bottom of the screen. Koy begins singing, her trilling contralto reverberating through the pines—Pan's beloved Echo.

I'm exhausted and crawl into my tent and drift off among irresolvable harmonics.

Morning comes very quickly. The sun is just rising, the forest mist just lifting when the mushroom hunters and I rouse ourselves from our sleeping bags and emerge stiffly from our tents. Koy has already been up, revived the smoldering ashes from last night's fire, and is preparing breakfast—a steaming Thai soup seasoned with chili peppers, spice powders, and chopped scallions. Everyone has soup, sticky rice, and cooks what pieces of chicken and beef remain from last night's dinner over the fire. As we eat we walk around to warm ourselves.

"We decided last night," says Koy, "that after this week we're going to Montana."

"It's better there?" I ask.

"I talked to my friend last night and she says there's more mushrooms and more places to pick. It's very slow here. It's too warm and so we have to climb too high to find the mushrooms."

They, too, are being chased by summer.

I tell her that I'm heading up that way, to see the glaciers in Glacier Park, so maybe we'll meet up there. The forests that were burned surround the national park.

"Do you have a map?" she asks.

I go to the car and bring her the map.

"Here is the best route."

She marks the roads with a pen.

The men have doused the morning fire and covered it with dirt. They pack the van and their pickups with collecting buckets, pocket their lunch bags of sticky rice and cooked chicken, and one by one drive off.

I thank Koy, we hug, and I tell her I'll look for her in Montana.

With the camp deserted and the sun now up I stand and savor the first warmth of the morning. I pack up my tent and sleeping bag and put them in the trunk but as I do I see that I've got a flat left rear tire. I pull out the jack, find a flat rock to balance it on, and change the tire for the small spare, drive off, get about five miles down the road and the spare goes flat. So I open the trunk, pull out the first flat, and stand with it on the shoulder of the road, trying to hitch a ride the couple of miles to Sisters. This is no empty rural road but no one stops until a white van pulls over. The door opens and who but Sam jumps out. He puts the tire in the van and I join the rest of the group inside, Koy, her husband, and Charlie.

They had planned to go north but changed their minds and decided instead to head to the mountains south of Bend. Lucky for me.

They drive me into town and leave me at a service station whose attendants give me strange looks when I come out of the van with Sam and Koy. We all say good-bye once more and they drive off.

With the tire fixed and the purchase of a used tire for a spare, the station gives me a ride to my car and soon I'm back on my way north.

As I drive I try piecing together a future narrative:

The original forest here was a pretty hardy mixture of ponderosa pine, Douglas fir, and western larch, huge trees beneath whose vaulted canopies little could grow. The pioneers could drive wagons and cattle through them. Nearly all of these trees were cut down in the last 150 years and the forest that grew back was a mixture of these, less hardy pine species, and undergrowth that can sometimes be hard to walk through. In the summer, when there is less water, plants photosynthesize less and tree growth slows. This is why tree growth rings provide such accurate histories of climate. In this mixed forest, heat and drought become far more stressful and stress makes a forest insect-prone, disease-prone, and fire-prone.

Over the last 100 years the average annual temperature in these mountains increased by 3 degrees Fahrenheit. Over the next 100 years the temperature will go up another 5 degrees. While the amount of precipitation should stay about the same, the expectation is that here, as in the Rockies, warmer winters will mean more rain—which will wash away quickly—and less snow. Much less snow. Over the next 50 years the snowpack in the Cascades will be half what it is now. By 2090, the snowpack in March will be what it is now, in May. Earlier spring and earlier thaw will mean less water in summer—less water and higher temperatures when the mountain evergreens are already under stress.

Example? In the mountains here, as in the Rockies, more than half the whitebark pine that once dominated the forest has been infected with a fungus, white pine blister rust. More than half of the trees it infected in Idaho are dead. And the infection rate in Oregon is more than 20 percent. Another? An epidemic of bark beetles is killing millions of trees from Arizona to Alaska.

There have been other long droughts, but none with temperatures so high for so long. The higher temperatures allow populations of insects that are usually confined to the South to move north. The heat also improves insects' sex lives. Some species of bark beetles now produce a new generation every year instead of every two years. During a particularly persistent bark beetle infestation in the 1930s a forest ranger composed the following reel:

The bugs they're killing the timber
They've worked for many a year,
But the Entomologists come and prophesy
That they'll quickly disappear.

Chorus:
Oh! They ain't gwine fly no more, no more,
They ain't gwine fly no more.

But how in the hell can the bug men tell
They ain't gwine fly no more?

The crews are cutting the timber
The crews are peeling the bark,
The bug men say the beetles they'll slay
And clean them from the Park.

After all was said and done, all that was tried, failed. The lesson learned, wrote a researcher recently, "was that once a mountain pine beetle population erupts over a large area of susceptible forest type, and as long as environmental conditions remain favorable, there really is no way to stop it until almost all the susceptible trees are either killed or removed by logging."

Perhaps environmental conditions won't get so bad. If spring begins weeks earlier, enough moisture may accumulate in the soil in winter to remain sufficient for the summer. Since CO_2 levels will continue to increase, perhaps tree growth will persist as the Duke studies showed and the growth won't deplete yet more soil moisture or nutrients. Trees can be resilient. Since they die naturally or are cut every 40 to 70 years, long-term climate change may not make a difference. But then the young trees may not be as resilient and they'll be growing in a different climate than the ones now ready to be cut. With the climate changed, what will grow back? No one seems certain but everyone seems edgy with expectations.

May 27
15 HOURS, 34 MINUTES OF DAYLIGHT

This has been a very lonely run, a long drive north across the 45th Parallel (halfway between the equator and the North Pole) in a fierce thunderstorm beneath a truly frightening sky, and on to the Columbia Basin, a plateau full of odd lumps and gullies, like cats make under a bedspread. This great basin resulted from periods of cataclysmic volcanic eruptions that began 17 million years ago and continued for 10 million years, burying whatever the geology once was under 6,000 feet of lava. Some 2 million years ago ice sheets began descending from the north. By 20,000 years ago the ice had dammed the Clark Fork River, which flowed out of the Rockies, creating a lake in western Montana 200 miles long and 2,000 feet deep that held 500 trillion gallons of water, more than Lakes Erie and Ontario combined. Hard as it is to imagine such a lake, imagine that when the ice dam gave way all of that water poured out within 48 hours, which, give or take a trillion, comes to some 10 trillion gallons an hour moving at 65 miles per hour, 10 times the combined flow of all the rivers in the world. The surge dredged the Columbia Basin to its bedrock, carried 200-ton Rocky Mountain boulders to the Cascades, poured across the Willamette Valley, and emptied into the Pacific. And this dam building, bursting, and brutal scouring happened not once but dozens of times over the last 2,500 years of the ice age.

I suppose the basin—so scarred by furrows and roughed up into gravel mounds, coulees, and camel-humped loess dunes that it is known here as the Scablands—has earned its present quietude. Denied any landmarks to give me a sense of perspective, I can't tell what is distant and what is close. I can see thunderstorms pouring rain maybe a few miles, maybe a hundred miles, away.

I decide I'd like to sleep outside here so come dusk I take a back road just south of Spokane, a road that appears to have no outlet, that winds among high smooth-domed hills so that in the twilight I feel as if I'm driving through a sea of high rolling swells and just as if I'm out on the sea I'm struck by a remarkable fatigue. I reach the campground that the map showed was nearby but it is closed, gated, and locked. The road does nothing but go on, rising up beneath me then swirling away behind for what seems like hours. There is only blackness in the rearview mirror; ahead, only the soft outlines of more hills. I'm afraid to take my eyes off the road even long enough to look down at the dashboard clock. I see a farmhouse across a flat field. And finally, I reach a main road and a motel, an old one-story ranch-style motel with painted green doors along a cement sidewalk. The room has a bed, a TV with a wire antenna, a washcloth, a small thin towel, a packaged bar of soap, and a funereal chill. Enough light shines through the curtains from the parking lot that I don't have to turn them on inside. I lie down on the bed. I've no interest in a drink.

For the first time on this trip I have a sense of my isolation from all but the nature of things. How long I've been in this state I don't know. Perhaps it began at the spring of the Melior. I recall that when my son was maybe three years old I took him out to New York Harbor to show him all the boats. He saw one of the ferries drifting in to dock and wanted to go on. We took the next one out and stood on the deck while we moved across the harbor. But when he saw another ferry pass us he became upset, pointing at it, saying he wanted me to take him out on the boat. I finally realized that he didn't understand that he was on a boat. It is like looking out into the night sky and thinking how wonderful it would be to be out on one of those planets drifting in space. Is the question, Are we where we think we are?

John Burroughs, the son of a Calvinist minister who decided to forgo religious investiture for a sharp pocketknife and good boots, wrote that "We love to fancy that we see friendly hands and hear friendly voices in nature ... to make ourselves believe that the rains, the warmth, the fruitful seasons, are sent by some Being for our especial benefit ... that there is air because we have lungs, and water because we need it to drink, and light because we need it to see."

But it is just not so, no more than we have legs because there are pants.

"In the universe, as science reveals it to us, sensitive souls experience the cosmic chill." Maybe that is what I'm feeling in this room.

Signs and portents are like prisoners of war: If you torture them long enough, they'll tell you anything. In the end, as Burroughs points out, there is no comfort save that of knowing things as they are.

"I have positively appeared," wrote Walt Whitman. "That is enough."

Maybe we appear and maybe we don't. These last few days I don't seem to find myself anywhere among the wide sky and senseless landscape. I wonder about this wandering. Maybe it is just a disappearing act. Something of a masquerade. Is trying to understand how the world works a way of getting away from the world itself? What is all this seeking and finding? Something that has no end, that is certain. It is not a search. The root of "seek" is to perceive, track down—same root as "sagacious" and "exegesis." The root of "search" means circular, a crown, a ring, to go around. A search, to me, is seeking of the worst kind.

Out in the spring, worlds large, medium, small, microscopic, and molecular link and unlink, pass around and through one another like magician's rings or jazz improvisations and the trick, if you have any hope of keeping track, is to follow the energy flow. A friend with whom I go canoeing claims that paddling insults the gods who represent the river current's benevolence. By his way of thinking, a reverential, moral, and compassionate canoeist paddles as little as possible. He himself measures his own state of grace by how many cigars he gets to smoke on the trip downstream.

Maybe I just need a drink after all.

May 28
15 HOURS, 36 MINUTES OF DAYLIGHT

The next morning I get out of the basin of ten thousand doubts as quickly as I can and head back to the forest. Koy said the best way up to Glacier National Park is to take Route 200 east and then cut off on little Route 28, which winds north across the Cabinet Mountains, to Route 93 to Flathead Lake. From there it is a straight shot to Kalispell and the Flathead Forests. This should be easy but after a day's hard drive in which I get past Thompson's Falls, I decide to pull into the state forest to find a place to camp for the night. The moment I do, I know—being the fool that I am for beauty—that I've found yet another paradise.

The narrow unpaved river road through the forest follows the course of the Thompson River, a 55-mile-long tributary of the Clark Fork, the river that roughed up the Columbia Basin 20,000 years ago. The Thompson is full of spring flow and afternoon light, running green and gold, sending cold spray across wide gravel bars. I forget I'm in a big white Chevy on this narrowing dirt road and keep heading upstream, taking small cutoffs into the woods as the one road intertwines with another, joins a wide logging road, then diverges again, until I'm completely lost and trying to navigate by a sun that has vanished beneath the mountains and the high pines. Somehow I've been able to keep beside the river whose current in the early evening rushes like thick braids of wire gray cables being pulled down a ravine.

Just before dark I stop to camp along what from my map looks like Plum Creek, a tributary of the Thompson, find a flat promontory above the stream bank, and set up my tent. Because of the wind, I dig a small pit, build a low fire in it, and make myself some tea and soup to have with bread and my staple can of sardines. God knows between the sounds of river and wind it is not quiet. But if Sam was right it won't rain.

At 9:15 it's still light enough to read by the light slipping through the flailing canopy. But it is cold just sitting so to warm up I put on a coat and cap and take a walk along the forest road. I imagine I hear all kinds of things—another car, a crazed hiker, a grizzly bear. My mind is no amateur at coming up with ways to keep me from moving on. It wonders why I'm getting myself mixed up with all this seeking and finding, this listening to wind and water, things that have no beginnings and end who knows where or when.

And then I see the elk, a huge elk, or maybe it just seems huge to me since I've never been that close to one before and it is standing, large as life, in a twilit clearing just two low shrubs away from me. The wind keeps him from catching my scent and the thrashing forest canopy and rushing water keep him from hearing my steps or the slight gasp I know I let out when I came upon him. His shoulders are high as my head and his head is huge and crowned with velveted antlers. He is pure presence in a fantastically misted glade. And then, when I can take my eyes off him, I see two other, smaller elk browsing among the trees nearby. Males, perhaps, that graze the forests in the spring when the females go off to calve.

As I stand there my mind starts in on me again, deflecting oncoming reflections with information. Fact: Now found almost exclusively in the West, elk once roamed across most of the country. Fact: Hunting and the cutting of the forests did them in but now milder winters with less snow enable more of them to survive to spring. Fact: Their new problem may be their increasing numbers. Lack of food may leave many winter survivors weak and unfit.

Why do I need to know *everything*? Perhaps I feel that what I know is mine alone, that it can't be taken away and I can't be separated from it. Knowledge is the mother lode. Give it up at your own emotional peril.

I'm falling for this elk.

My mind tries another tack: "Maybe you're not here at all. Maybe someone is dreaming of you meeting an elk in the forest. Maybe you're hallucinating. Maybe the elk is hallucinating you. How would you *know*?"

I remain still in this moment's moment. I shut up my mind so there is no question of my presence. I am positively here: no dream, no panic, and no hallucination, just a giddy infatuation with an elk.

I walk back to my tent and, taking Sam's word, pull my sleeping bag out, slide in, pour my nightly whiskey, put on my headlamp, and look at the map, which seems to show a road north and over the mountain. I say "seems" because I'm looking at a road map and not a topo or trail map since the ranger station was closed as I came through. It is one of those roads with dashes outlining it, which means it might be closed. But going back would mean a long detour east. It would also mean remembering all the turns I made to get here, wherever this is. I decide to try to make it over the mountain.

The morning is cold but dry. I pack up and head out, aiming north. When the road leaves the river it begins to head up the mountain in switchbacks carved out of the rock. Soon it is down to dirt and barely one lane, just wide enough for the car but pretty scary since there is no guardrail and the slope is steep. My mind worries me over the possibility of a rockfall and this time I think it's got a point for there can't be more than a foot or two of road between the car's tires and Kingdom Come. Worse is the realization that there is no turning back because there is no place wide enough to turn around.

As I drive higher the pines thin out and the forest bears all the marks of one that was logged. The bad news is that the logging must have caused a tremendous run of silt down into the river below, which is why here as in other Western streams and rivers the native bull trout is threatened or endangered. The bull trout is actually a salmon, a char that migrates from a lake or one place in a stream to spawn in among light loose gravel. Its eggs need a great deal of aeration and the silt from road construction and logging pretty much does them in quickly. Unlike sea-run salmon they don't migrate in spring but wait until the spring waters go down and the current slows to make their runs in June or July.

The good news is that if logging trucks used the road I assume it is pretty solid. Or at least was at one time. Who knows what a winter of freezing and a quick spring thaw could do to it. But higher still I see new cuts so the logging must be pretty recent.

I'd love to get out to look around but first of all I'm afraid to stop for fear that the Chevy might not start or might spin its wheels on the wet dirt, trying to move up the steep slope from a standing start. Second, I don't have enough room to open the car door. If I get out the passenger side I'll drop probably 6,000 feet. So as I drive I try to take quick looks out. The view is stunning. The chasm is bottomless. The river runs somewhere below, angling through the steep forested slopes. Looking at the peaks across the gap I can tell I'm nearing the summit of this mountain and with luck the road will continue over and down the other side.

And then hope vanishes. Coming down the road in front of me is a huge logging truck with rough-cut 30-foot sections of pine stacked up and strapped to its bed. I stop. The logger stops. He must think he's hallucinating. I'm sure he's met up with Jeeps and Ram Toughs on this road, but the sight of a white 1984 Chevy Impala probably puts his willing suspension of disbelief to the test.

He opens his window. I open mine.

"G'morning," I shout.

"Yeh. G'morning," he says. "How did you get up here?"

I think it's a polite way of saying, "What the hell are you doing up here?"

He opens his door, squeezes out of the cab, and takes a few steps toward the car so we don't have to shout at each other. Impolite as it is, I've got to stay put.

"I came up from the river. I was hoping this road would take me over the mountain."

"No," he says, shaking his head with some puzzlement. "It won't."

"Oh."

It's hard to get a good look at him since the Chevy's windshield has a bit of a fun house effect on things up close. He is not too broad a guy, maybe in his 40s, dark eyes, unshaven. He squints one eye at the car.

"It's closed up on top."

How closed if he is up here? I haven't seen any snow on the surrounding mountaintops. Although this may be his first load of the spring, when trees are cut before the sap runs up their trunks. The road above could be pretty muddy. I've got no choice. In the end he is the gatekeeper.

"I guess I have to go back down."

"Yeah." He turns for a look back up the slope.

"Is there a place to turn around?" I ask.

"I can back up a bit. There's a pullout."

"That would be great. Thanks."

He climbs back into the cab, grinds his truck into gear, turns his wheels, and then slowly backs up about 20 feet, until his cab couldn't get any closer to the road's edge. As I pull around him I stop to thank him again, but before I can he peers down and says, "You drove that car all the way up here?"

I don't know whether to smile and have him think me a grinning idiot, or just nod and have him think me a mindless idiot.

"He's pretty tough," I say, slapping the car door, to which the logger just shakes his head. "When I get down, is there a road out of here?"

"Just stay on the log road, the gravel road. It'll get you out."

He gears up and goes on his way and I feel better knowing that this big truck is able to maneuver these grisly switchbacks. *So long as your brakes work,* I decide, *you'll probably be fine.* I pull up and as he said, there is a cut in the cliff for a pullout. I turn in and sit, giving my heart a moment of calm. I get out of the car to stand and take in the view of the dense forests once more. The river's mist is rising up through the gorge. I walk up the road a little and while clear of snow I can see it becoming increasingly damp and slick. I come back to the car, gauge the radius in which I have to turn around, get back in, and after a few reverses I'm headed down.

I follow the logger's tracks, and automatic transmission or not, I downshift into first in the hope of keeping the car from speeding up. I check the brakes before every curve and focus all my thoughts on the car and the road in order to keep from imagining what might happen if the brakes don't hold.

All in all, I think when I get back to the river and flat ground, a fine place to have gone finding.

7

$\mathcal{M}$ushroom $\mathcal{H}$eaven

15 HOURS, 43 MINUTES OF DAYLIGHT

It's a straight shot of about 2,500 miles from Louisiana to Montana and I've managed, through a highly discerning sense of directionlessness, to make a trip of 6,000 miles out of it and arrive in Glacier National Park just as the season is getting decked out in its fluvial finery. Glacial ice sports its powder blue luster. Streams flash fresh snowmelt. The hourglass flips and time begins to run on the short-lived northern spring. Tourist season hasn't arrived so the woman in the rental office gives me a good rate on a little cabin in the park, explains with practiced exactitude how to turn on the pilot in the hot water tank—I nod without comprehending a bit of it—and hands me a large skeleton key.

My cabin stands ensconced amid red cedar and western hemlock and just a couple hundred feet from Lake McDonald, the park's largest body of water. From Lake McDonald's shore I can look east and see the 9,000-foot peaks of the Continental Divide—at least I can when it is clear, which is seldom. Most of the time the mountains rise into an eclipsing mist. The mist and the lake seem to be in a continual exchange of damp spirits.

My own spirits are fine. This is my last and most northern stop in the Lower 48, along the 49th Parallel, which divides the United States and Canada. With three weeks of spring remaining, the light lingers until just after 9 P.M. Since my trip began in March, I've gained some four hours of daylight—just a sixth of a day, but with these extra four hours the average amount of solar radiation reaching the northern hemisphere each day has more than doubled. The actual amount depends upon where you are. Today in Louisiana, for instance, daylight will last an hour less than it does in Montana. Spring in the north is brief but intense.

With a couple of weeks left before I leave for the Arctic, I decide to remain here and explore what was once the greatest collection of glaciers in the country. In the Park Service office nearby the cabin, a ranger in his formal Park Service greens stands by a scale model of the park. It reminds me of the model train set I had when I was 10 that my father set up on the Ping-Pong table. This was not one of your H-O scale setups, this was heavy metal Lionel with a black Bakelite transformer the size of a table radio that, if turned up to full, could run the big engine and all its cars straight off the track.

A troop of Boy Scouts in their own greens and red neck scarves stand around the park model, more interested in the miniature world in front of them than in the ranger's presentation. For some reason I now find these kinds of models unsettling. Maybe it is their deadly immutability. A world that can gather dust. Glacier is not that kind of a world.

For a million and a half years, beginning about 2 million years ago, continual cold gripped these mountains. Ice sheets thousands of feet thick advanced and retreated, gouged the mountain rock, and left behind grooves, scrapes, and hollows that wind, rain, and the occasional arid millennium eroded further. What life migrated here was beat back harshly and often. For most of the era all but the highest peaks stood above a frozen plain of dazzling albedo: the sun stared down and got an icy stare back.

Some 12,000 years ago the great ice sheets began to recede once again and for the next few thousand years a kind of continual spring reigned. Bare rock that had been hacked at unmercifully was quickly reshaped by the life that came and colonized, at first tentatively, and finally as vigorously as the warmth and water allowed. Human remains have been found that date back 10,000 years. By 6,000 years ago all the glaciers had vanished. Then, a couple of thousand years later, the cold climate returned; winter snows remained throughout the year and began to accumulate. Buried under the thickening snow, snowflakes turned to ice crystals, and like clothing in an overstuffed suitcase, the squeezed crystals stretched to fill up the air pockets in the snowpack, which then became increasingly dense. Under intense pressure the bottom layers of ice turned plastic, fluid, and the huge masses of snow cradled between the high peaks began to move, irresistible rivers of ice dredging their own gorges.

Unfortunately, I've arrived in Glacier 150 years too late to see these majestic ice floes. On my first foray into the park along a startlingly swift stream running with spring meltwater that appears to be pouring out of every rock along its banks, and even surging up from out of the ground, the glaciers are more conspicuous by their absence. Steep valleys once slung with ice now appear as emergent meadows. The glaciers I do see lie slumped in their mountain hammocks, snowfields withering in the Holocene spring.

If one could pick a date when the most recent melting began here in Glacier's high peaks, it wouldn't be in some distant geologic past. In fact, 1850 might be a good choice, for sometime around then all of the park's 150 glaciers began to recede. A glacier melts like a block of ice: the sides melt away as the center thins out. In Glacier this melting back has averaged some 20 to 60 feet a year. It has not been regular. Between 1966 and 1979 several glaciers even made hopeful advances. Since 1979, however, the melting has accelerated. Only 35 glaciers remain and even the largest of them, Grinnell Glacier (named for George Bird Grinnell, a journalist and big-game hunting companion of Teddy Roosevelt's who, in 1900, advocated for Glacier's park status), hardly attains glacier status. In 1850, Grinnell Glacier covered 575 acres, about one square mile. By 1960 its area was reduced to 315 acres, and by 1993, to 217 acres. Only 10 percent now remains and the expectation is that in another 25 years Grinnell, and all the rest of the park's glaciers, will vanish.

I have no nostalgia for the ice age. Glaciers have come and gone often enough and besides, the scenery they leave behind is not only spectacular but also more accessible once they're gone. But this going is going *fast*. And not only here. Since 1850 most of the world's midlatitude glaciers have been receding—in the Andes, Himalayas, Alps, on Mount Kilimanjaro in Africa—just as quickly. What is notable about the retreat is that lack of snowfall (although in many places that is occurring) is not the primary cause. For glaciers to retreat as much as they have over the last two centuries either snowfall would have had to decline by 25 percent or average temperatures would have had to increase by at least 0.5 degree Celsius. Since there is nowhere that precipitation has decreased by anything like 25 percent, glacial retreat confirms the increased rate of global warming—for those who might still doubt it. Less precipitation will exacerbate the retreat since every degree Celsius increase in average summer temperature will melt approximately a meter's thickness of ice.

June 2

15 HOURS, 45 MINUTES OF DAYLIGHT

Not long after I begin my trek along Glacier's Avalanche Creek on a trail that meanders for a couple of miles through old-growth red cedars, I hear the insistent bell of a bear-wary hiker up ahead. During the summer this is an especially crowded trail but in spring, when cold moisture fills the forest with a chill mist, there is no one else but the bell-ringer. Being aware of bears is not optional when hiking but declaring one's presence by ringing a little bell nearly every minute, while often recommended, has on me the same effect as a cell phone going off in a concert hall. I've heard all kinds of ways to keep grizzlies at bay or, should none of them work, how to fend them off—everything from firing pepper spray to playing dead—but a guide I know who camps often in grizzly territory says that just by being aware and keeping a respectful distance he has never had occasion to use the gun he carries with him. He also recommends against wearing brightly colored backpacks—bears are curious—and that if a bear shows up in your campsite bang a pan. How much good pepper spray or even a gun will do may be questionable since out of the blocks an otherwise lumbering bear is moving at 35 miles an hour.

Because most of us are satisfied with the illusion of safety, I guess a bell is as good as anything. I recall reading that sailors in the Pacific during WWII were issued, in case of having to abandon ship, a shark repellent to spread around them in the water. The truth was that the navy knew the stuff was useless but issued it anyway, figuring, I guess, that if you didn't get eaten you wouldn't complain and if you did, well, you wouldn't complain either. I try to walk through the woods very aware even when looking for mushrooms or wildflowers but I know that sometimes I become so interested in what I'm seeing that I forget about bears and time—both of which, if given the chance, can gnaw you to death.

What I decide today is to let the bear watchman with his bell get far enough ahead of me so I don't hear him and then take a slow walk through these extraordinary woods, among trees 10 times older than I. This forest with its high spacious vault and nearly enclosed canopy has lived here for 500 years—15 paltry human generations come and gone—slowly creating the conditions that suit its survival. Growing in shallow soil the trees' own fallen branches and needles provide a thick fertile base, one that might invite invading plants, but few besides ferns and mosses can gain a foothold beneath the trees' dense canopy. Even when the summer heat and drought dry up many plants outside the forest, within, the deep shade remains sodden. As I walk I find myself chilled and sweating at the same time.

Spring flow floods the creek. The walk takes me upstream, where the gorge narrows and in places the stream cascades down 20-foot cataracts, fills flat gravel washes with turbulent foam, swirls up over its banks, and then rushes even more swiftly through rock-choked chutes. As the trail leaves the stream bank I'm amazed that I don't get far before the sound of the water disappears, absorbed by the soft surfaces and dead air of the forest. The trail rises a few hundred feet and then meets up again with the stream, this time where it pools in a cold little marsh, just before the Lost World vision that is Avalanche Lake, a misty basin filled with cold glacial meltwater that streams out of the face of a high cliff wall. The walk is too easy, I think, to end in such an extraordinary place.

In the lake drift a small flock of goldeneye ducks in spring plumage—deep purple and bright ochre. The bell-ringer, a man with a walking stick, is just heading back. A young couple ask me to take their picture, explaining that they're here on their honeymoon. They seem very young.

"How old are you?" I ask.

"I'm twenty and she's nineteen," he says.

"Well, congratulations," I say, but like the father that I am, I wonder about them being married so young. Then, of course, I was married at 24, which is not so much later, and human time, as this forest attests, is pretty mean. And it is spring in their young lives so why shouldn't they be together here in this Dionysian domain (which Plutarch described as "the whole of the *hygra physis*," all of "moist nature"), a forest that in its wholeness has lasted 500 years. Human wholeness is not that durable.

This may be what nature inspires in spring, a longing for

wholeness free from time's constraints. And perhaps this longing, catalyzed by histamines and hormones, is what in spring turns us lightly to thoughts of love. Dionysus's minions were all about joining nature and Eros. As the Athenian playwright Euripides saw them, they could strike a rock and out would gush streams of water:

> *Another hurled her staff against the ground,*
> *Whence the god sent a wine-spirit bubbling up;*
> *And anyone who wanted white drink, needed*
> *Only to scrape the ground with fingertips,*
> *And milk came spurting; from their ivy-clustered*
> *Thyrsuses dribbled down sweet streams of honey.*

Perhaps my spring chase has been all along about my own longing to reacquaint myself with such wholeness and enjoy its rewards, to exorcise whatever mind it was that had become wholly tied to time and its civilities, to seeking or to finding. But I'm thinking now that seeking and finding is one endeavor, not two. Perhaps I should have taken this trip sooner. The 4th-century Chinese poet T'ao Ch'ien wrote, "A heart that is distant creates a wilderness round it." Maybe there is nothing better than surgery to clear away the accumulated brush. If this couple weren't already married I'd perform the ceremony right here for them myself.

June 3
15 HOURS, 46 MINUTES OF DAYLIGHT

No conversation in the little towns around Glacier is complete without someone saying that they know exactly who is to blame for the devastating summer forest fires—it's usually the Park Service, Forest Service, environmentalists, or politicians from the city or the East. The repeatable euphemisms are "Big Brother," "tree-huggers," and "New Yorkers," all of whom, the argument goes, have put the loggers out of business and in so doing allowed the forests to become ripe for burning.

One woman, a volunteer in the Forest Service office from whom I was asking directions, became very nearly apoplectic when I asked her about the fires.

"People come out here from the city, that's fine. But don't try to change the way we live! You say fire is just natural? I say, 'Okay, Mr. and Mrs. Green, you like fire? Let's see how much you like it when it's about to burn down your house!'"

Not knowing how I touched off this explosive harangue I thought I'd cajole her a bit by saying that I was sure the Forest and Park Services had the forest's best interest at heart. This, however, provoked a gimlet-eyed stare and another firecracker burst.

"The Park Service! The Park Service doesn't allow this, doesn't allow that. They've made the park into a research station and they won't be happy until every animal has a tag on its ear. Well, why don't they come and try to tag the bears that now come around all our homes? We can't shoot them anymore!"

I nodded, smiled wanly, thanked her, and walked away.

Last summer some 737,000 acres of Montana's forests burned. It was not the record million acres of a few years ago, but it was more than you'd like. In Glacier alone 136,000 acres burned: that's 212 square miles, about 12 percent of the park and the largest fire in the park's 93-year history.

Mitch Burgard, the park's assistant fire management officer, had offered to accompany me out into the burn area so I called him and we headed out in his official green truck. Mitch is a lanky, laconic, well-focused guy who, at least today, won't argue about fire with anyone.

"Fire is not good or bad. It's part of the life cycle of the forest. What you had before you're going to have again."

We pull over by burned stands of lodgepole pine above the north shore of Lake McDonald. The remains of trees stand like burned matchsticks. Unlike the forest in Oregon, there is far less bare rock around, the ground is black and flat. Mitch bends down and scans the charred forest floor.

"The cones of lodgepole pines," he says, "can only release their seeds under heat, 700-degree heat. When a lodgepole pine forest burns rather than being cut down, it assures that another lodgepole pine forest will grow in its place.

"A lodgepole pine stand burns about every hundred eighty years. When we see it, it's a disaster. But our lives aren't long enough to grasp what a hundred eighty-year interval is. If winter were a hundred eighty years long we'd think differently about spring. Well, this forest will be in its spring phase for the next thirty to forty years."

"What about the forest animals?" I ask.

"The animals are well enough adapted. We never see a roasted carcass. When the fire comes they move. During the Moose Fire when 70,000 acres burned we saw a mass migration of boreal toads getting out of the way."

Mitch is sticking to his story. Fires have burned the forest, fires will burn the forest. The forest needs fire.

"You see," he says, pushing two fingers into the ground, "there's only two or three inches of litter on the forest floor and so long as the fire wasn't hot enough to cook the underground rhizomes, the new growth will have everything going for it. No fighting over moisture and sunlight."

When he pulls out his fingers the tips are not blackened but gray from the less burned ash beneath. Under the winter snowpack, he explains, the burned soil decomposes further. The spring meltwater then adds its share of nutrients.

Bending down next to him I can see green shoots are already rising. He identifies them: "Fireweed, rose, bear grass, willow, and look—all those are lodgepole seedlings."

Miniature trees, a couple of inches high, stand staggered across the black soil.

"Bicknell's geranium," he continues, "is a plant that *only* shows up after fires. After three or four years they disappear. Then there'll be huckleberries, which will make this great bear habitat."

But suddenly I'm only half listening because goddamn it, the forest floor is carpeted with more morel mushrooms *than I've seen in my entire life.* And they're huge. Black or gray as ash and with heads as big as pinecones. Thousands of pungent Dionysian thyrsuses. I'm trying to listen to Mitch but in my head I'm counting basketfuls, pounds, I'm stuffing big morels with fresh crabmeat, serving them with pasta, with spring lamb and fiddlehead ferns. Mitch sees me looking at them.

"Pretty amazing, aren't they?" He smiles.

I shake my head. "I've never seen anything like it." Koy and her crew, I think, certainly knew what they were talking about.

"Too bad," Mitch says, "there's no picking in the park."

I nod numbly.

"Research," he says.

I'm thinking, *Big Brother. Tree-huggers. Damn New Yorkers.*

There are, however, some 350,000 acres of just as wonderfully charred national forests surrounding Glacier, and you can pick in them all. Tomorrow, I decide, I'm off mushrooming.

June 4
15 HOURS, 48 MINUTES OF DAYLIGHT

The weather is still lousy—the wet chill finally took occupancy of the cabin and I wake up in its clammy bed—but I'm still gripped by yesterday's vision of morels and so get up, make some coffee, throw on my rain gear, get into the Chevy, and head off to buy a mushroom permit at the Forest Service office, where I'm greeted by the gimlet-eyed woman with whom I spoke the other day. She doesn't seem to recognize me.

"I don't know," I say, "if I'm going to be selling any but just in case I strike it rich I thought I'd better have a permit."

"You better get a map, too," she says. "Which forest are you going to?"

"I don't know," I say, but the eye quickly makes me realize this was a mistake. "Where have they been working?"

"They're everywhere. The Asians came in this week."

"Where are they camped?"

"They're camped everywhere. There's probably 2,000 of them. They get more every year."

"Where are the buyers located?"

"You've got them north and south."

Things didn't get better when I handed her my New York driver's license. That eye—I recall my grandmother refusing to allow one of my mother's friends to visit my brother in his crib because "that woman," she said, "has the evil eye." And she spit between her fingers. When my brother needed eyeglasses at five my grandmother said she knew that over her objections my mother had let that woman in to see my brother, the woman had complimented my brother's eyes, and no one had said a *keyn 'eyn-hora,* which in Yiddish means "no evil eye" and serves to reverse the curse of a compliment from someone with an evil eye.

"We don't see too many pickers from New York," she said, staring longer than I liked at my license.

"I'm just passing through."

"Well, here are the maps and here are the rules. No picking in wilderness areas. No picking in closed areas. No driving on closed roads."

"Thanks," I said.

"Well, good luck."

And taking a last look at her—or her at me—as I left, I found myself involuntarily uttering *keyn 'eyn-hora,* then wondering if the expression had ever been used with regard to mushroom picking. Why not? Picking mushrooms in Eastern Europe was and remains a ubiquitous pursuit. Only in America, where Puritans inspired a fear of everything in the dark woods, do we suffer from myco-phobia. In all the years I've been out in the forests of Pennsylvania, New York, and New Jersey picking mushrooms I've never once run into anyone else out picking. In Eastern Europe if you aren't out in the forests before midmorning, everything has already been picked.

With the curse reversed I head off for the burned forest. The acreage available is huge and most of it lies south of Glacier Park on either side of the South Branch Flathead River that runs through Flathead National Forest. Since no roads cross the river I have to choose one side and with no reason in mind I head down along the east road. The forests rise steeply from the right side of the road and drop off just as steeply to the left, down toward the river. Everything has been blackened but with morels in mind my concern for the lost trees gives way to anticipation of the crop that grows this spring among the ruins, sucking away at the caramelized nutrients beneath the charred soil and produc-ing prick-capped fruits above. Besides, I'm going to assume, as Mitch Burgard said, that this is the forest's first spring in 180 years. This is shroomer fever, no longer a friendly walk in the woods. Every picker for himself, the territorial imperative takes over.

I see plenty of evidence of "the others." Their pickup trucks pulled off onto the shoulder of the road tell me where they are now; tire tracks and cigarette butts tell me where they've been. Since I can't wait to start I pull over at a spot where there is a steep drop to the river. Pickers have been parked here, but since there is more forest above than below, my guess is that they went seeking upslope and not down.

I empty the contents of my backpack onto the backseat of the car and begin climbing down. The ground is black as pitch on the surface, ash and sand beneath. With no hangs to grab I've got to lean far into the hill and try to dig in my boots. Sideslipping the slope I find I'm gouging out the soil and probably some mush-rooms with it; the dust blows up in my face and coats my clothes. So I decide I'm better off just crawling along the ground. Morels demand that you meet them face-to-face. Soon, I see one and then another, and then more. Glacier's forest was far more flush but I'm picking plenty, enough so that I have mercy on the minia-tures and leave them to grow. Within an hour, stopping only to put my backpack up to my face to inhale the sweet must, I look as if I've been working in a coal mine but I've got probably four pounds of morels.

I work my way back up the slope, open the rear door of the Chevy, place a layer of paper towels on the ledge behind the back-seat, pour the mushrooms over the towels to dry, get into the car, and drive. A few miles down the road I see a rough hand-painted sign that reads, "Mushroom Camp," and I turn onto the gravel road. Tents have been set up along with fire pits but asking around, I find no one has seen Koy. Perhaps, they say, she is up north. I decide I'll head north tomorrow and just continue picking this part of the forest today.

A few miles farther down the road the river narrows and the forest spreads up into terraced slopes. I take a gravel road but soon find myself driving M. Dick up a hard-packed dirt road, dusty and potholed, that slings rocks up against doors. With the red soil run-ning in streams off the mountain I can see how quickly the forest soils erode once the trees are gone. The road soon ends in a muddy cul-de-sac, heavily scarred with tire tracks. I pull the Chevy over and get out.

This is one of those cold-hot mountain days, and in a forest struck by firestorms there is no shade to block the hazy sun, no trees to block the chill wind. I must be at about 5,000 feet and the temperature is probably only in the 40s. Where I've ended up is more damaged than anyplace I've seen. Avalanches of loose rock and burned trees must have come sliding down the moun-tain. I walk up along an eroding watercourse in which small wildflowers grow. In the mud I see the footprints of a bear who is made his way across the tire tracks and up into the woods. The prints are lined with rime and so I assume they haven't been made recently but nonetheless I keep an eye out as I walk. I find one morel quickly but don't see another for about 10 minutes. I assume I'm at the wrong elevation. The flushes begin low and move higher as the mountain soils warm. But with a good day's take drying in the car my fever for picking has cooled and instead I sit and admire the distant 8,000-foot summits to the south. These moments always come and they're always just as good, the moments when I arrive at a break in time, with neither the need to go on nor the urgency to go back, when the mise-en-scène is set—me, a large white Chevrolet sitting incongruously in the midst of a muddy forest road a few hundred feet below, and the vista of mountains beyond—but no one has the slightest idea of what happens next. It's a photograph that would defy captioning, a photograph you'd probably throw out for its lack of angle, drama, or perspective—"No life in it!" But such moments are precious, I think, for being free of the need for durability. Ecstasy, too, is fleeting and timeless.

Oh, this is very good. I pick up some wine on my way back to the cabin, grab a couple of fistfuls of morels from the back of the car, set a pot to boil for pasta, put butter and olive oil in my cast iron pan, add some chopped shallot, and toss in the dusted-up and sliced morels. The aroma of the mushrooms infuses the rising steam and replaces the cabin's chill with balmy vapors. The mushrooms give up their water, and when delicately deliquesced I add salt—always last—pour them over the drained pasta, add pepper, some heavy cream, a tablespoon of grated pecorino Romano (parmigiana's too fruity, but if I had a touch of fontina I'd add it), and serve with an Oregon cabernet. Spring is here.

June 7
15 HOURS, 51 MINUTES OF DAYLIGHT

After another day mushrooming and driving I was exhausted. I can't imagine what kind of fortitude the pros must have to pick every day for months at a time. I picked another few pounds and along the way passed plenty of pickups with California and Oregon plates but no white van. In one spot, as I was coming out of the forest and heading to the car, a pickup went by, the driver honked the horn, and his passengers in the back, all Asians, waved. I assumed they must have recognized the Chevy from Oregon. They are a very connected community and I thought how much they must know of the life of the Northwest Coast forests, for their livelihoods, like those of any nomad, depend upon recognizing the signs of the changing seasons, when to stay put and when to move on. Next month, when the weather becomes warm and dry, the pickers will remain in Montana to harvest forest huckleberries before moving south once again in September, back to the southern Oregon Cascades.

Yesterday I was too beat to move. I slept in, cooked a morel mushroom omelet, and then went to the store for some plastic bags to pack the dried morels, which were now withered to a quarter or less of their original size but whose aroma clings to the car's interior like pipe tobacco smoke. I read up on Alaska, pulled out my maps to plan my long drive to Seattle—where I'll leave the car and pick up my flight to Anchorage—and picked up the local paper, where I read that within the last month three Indian teenagers have died of alcohol poisoning.

When I left the East the papers and the TV news, following the death of a college student from drinking, had been full of stories on the subject. Universities began instituting no-alcohol parties, bars and liquor stores were given stern rebukes and warnings, and local congressmen spoke out. Meanwhile, these three Indian teenagers, not even out of high school, rated a two-inch article with no comments from anyone—just another report from the reservation.

Montana has much reservation land and the stories of all the tribes are pretty similar. The Blackfeet, for instance, had been pushed west by European colonization until they found themselves on the Western plains and finally in control of all these northern Montana lands and forests. They were nomadic bands that moved with the seasons. In spring, as the buffalo moved out from the forests to browse on the young grasses of the plains, the Indians followed. In summer they harvested berries. When the first White trappers appeared in the 18th century the Indians traded with them. When, in the 1850s, White settlers arrived, they fought with them. After the Civil War, as more settlers arrived, the Indians went from being admired as noble savages in a sublime Eden to being reviled as fearsome savages occupying the United States' Manifest frontier.

As environmental historian Mark David Spence writes in his book *Dispossessing the Wilderness,* even those who sought to preserve the Western wilderness wanted uninhabited wilderness and so promulgated the myth of the vanishing Indian at a time when the Indian cultures in places such as Glacier were still vital. They weren't vital for very long. Once the buffalo had been hunted close to extinction, the Blackfeet lost their main source of food. By 1882 the tribe was destitute and starving and dependent upon the U.S. government. To gain assistance they agreed to give up their land in eastern Montana in exchange for the tools to farm marginal lands. In 1885, the United States, seeking to mine for minerals in the Glacier mountains, again offered the Blackfeet assistance in exchange for their western lands that ran across the Continental Divide (that the Blackfeet called "the backbone of the world") west to the Flathead. Soon after, the railroad came across Blackfeet lands, and then Glacier was declared a national park.

There is plenty more to the saga. And there were some good guys—without the work of George Bird Grinnell (whose glacier is melting away) and the character of Woodrow Wilson, the Blackfeet might have been utterly dispossessed. But what struck me about the brevity of the newspaper article on the kids' deaths was the same thing that I felt when driving through reservations on my way north: The brochures and signs ("Photograph the Past!") and guides, for all of their now politically correct admissions that the Indians had existed and that they had been driven from their lands, still seem to treat the Indian nations in the past tense, as if they actually have already vanished.

I have a single plan for today and that is to drive over the ridge of the Rockies to the park's eastern slopes. The only road that will take me there is the winding 52-mile Going-to-the-Sun Road that bisects the 1,545 square miles of the park and crosses the Continental Divide at Logan's Pass, 6,646 miles up. Covered with snow all but a few months of the year, the road opened only a week ago when crews succeeded in clearing away the last of the winter snow and rockfalls.

The road is a narrow blacktop that climbs the mountains in switchbacks, the wall of the Rockies on one side, a low stone retaining wall on the other. It is a masterwork of engineering, as if someone had picked up the Great Wall of China and draped it over the Rockies. The sheer drops and winding climb remind me of my preposterous drive up the logging trail. This road, at least, is paved. But it is also two narrow lanes and the Great White takes up its full share, leaving little room for oncoming cars to pass. With solid ground beneath me at some steep rises I look out over the hood, see nothing but sky, and have the sensation of taking off. The road so twists around the mountains that if I didn't stop often at the provided pullouts I'd never get to appreciate the views, which run from the extraordinary to the spectacular. Most fascinating to me, however, is that I can see how the glaciers had distressed the land and shaped it into these rises, ridges, escarpments, and valleys, then withdrew, vanished, and let life take its course. And from the pullout at Logan's Pass with water from the rocks above coursing through culverts beneath the road and the sun in the big Montana sky gleaming on valley glaciers and broad meadows greening below and below them the coarse verdure of the vast forestlands and all of it streaked by blue streams that flash first here and then there, what I see in big-screen 3-D is life shaped by 10,000 springs.

*U*nder the *N*ever-*E*nding *S*un

June 15

19 HOURS, 19 MINUTES OF DAYLIGHT

With the Chevy sitting it out in the Seattle airport, I flew to Anchorage and this morning into Mudhole Smith's little airfield in Cordova, which is a fisherman's town on the eastern shore of Alaska's Prince William Sound. As the plane came in over the sound, I could see swirls of rich blue-green water, the reflected sheen of chlorophyll-laden phytoplankton, small marine plants such as algae that respond in spring to the increasing hours of daylight (back in Seattle today the sun sets four hours earlier than it does here), and the fresh spring runoff of glacial meltwaters flush with nutrients. These also bring the birds to the Copper River delta and to the tidal flats that surround Cordova. From mid-April to mid-May some million migrating shorebirds land each day in these wetlands, the most abundant being western sandpipers and dunlin, joined by dowitchers, least sandpipers, greater yellowlegs, and red-neck phalaropes—birds with thin bills, long legs, and knuckly knees—making their final refueling stop before they reach their Arctic breeding grounds. I hope to see them there.

Once the commuters who flew in with me have departed the roadhouse terminal I figure the hale balding fellow with the well-trimmed mustache who greets me is Paul Claus.

"Hey, Paul." And I put out my hand.

"Terry."

"I'm sorry."

"No, Terry. Terry Smith. Grumman Albatross?"

"Terry! Great to meet you. Is Paul here?"

"No, Paul's going to meet us up near his lodge. Terri and I will fly you there."

The Grumman Albatross! This is cool.

He grabs one of my bags and we walk over to the landing field.

"I'm only sorry we couldn't bring the Albatross," he says.

"Oh, no problem." But I feel a bit let down.

"Instead we brought the Goose."

Albatross, Otter, now Goose. "That's great," I say, but have no idea at all what a Goose is.

As we go through the gate and walk out on the tarmac I see that a Goose is not as big, but every bit as cool as an Albatross. I would love to have pulled up to it in the Chevy.

"There it is," he says, and waves to his wife, Terri, who's standing by the pontoon under the high wing of a sweet snub-nosed bumblebee, a minisub with wings. "The Aleutian Goose." Picture Bogie and Bacall coming out on the tarmac in Havana.

Grumman, Terry says, built the Goose before WWII. Designed to float on its broad belly (and bob gooselike in the water) it became the commuter plane of the wealthy, who could land it right on a resort's beach. The plane proved so versatile during the war that the navy commissioned a larger big-load-carrying model, the Albatross. Few were built and aside from Terry and Terri's, even fewer remain in use.

Terri has a sure handshake and flashy smile. She wears a khaki field vest over a light blue shirt and, wearing her dark aviator glasses with her auburn hair tied back loosely, she seems to me the very model of an aviatrix. I tell them both how much I appreciate their meeting me.

I get in behind the two of them and they give me a headset with a microphone attached so we can all converse easily as we fly. Terry starts the plane up and we seem to go hardly 100 yards down the runway before we're in the air. This is a pilot's, not a computer's, plane. Terry has to work it, manually adjusting the wing trim, steering, all the while his feet working the rudder pedals. We fly at 500 feet above the Copper River in a light rain, tracking north then east along the Chitina, a course that provides Alaskan panoramas of high glacier-draped mountains and broad river valleys.

Spring runoff has been heavy and sudden and the glacial streams, wavering stripes of aquamarine, twist their way down mountainsides of dense spruce forests before spilling out between high beds of sand and gravel. Below us is the continent's largest assemblage of glaciers and peaks above 16,000 feet—the Chugach, Wrangell, and St. Elias mountains—which converge in the 13-million-acre Wrangell–St. Elias National Park and Preserve. Soon Terry brings us down to barely 50 feet above the ground and we can see moose and buffalo hightail it from our shadow. Lifting above a screen of trees we come down on the other side and land on a grass meadow where a small plane sits—a very small plane—a plane that might be mistaken for a golf cart with wings.

We're greeted by Ben, a young man who works for the Clauses who I guess couldn't be older than 17. Since Terry and Terri will be heading north Ben is going to fly me over the Clauses' lodges. I thank Terry and Terri, tell them I hope I'll see them again. Terry starts up the plane while the three of us push the tail to turn it around so it can take off. Terri hops in and again with hardly a running start, they're off.

Ben and I squeeze my luggage into the little plane. I get in behind him—I've got to straddle the back of his seat as if I'm on a motorcycle, and my knees touch either side of the hull when I do—I buckle my seat belt, he guns the engine, and we're up. We clear the trees and rise into a sky that seems less vast than the land-scape of forests, mountains, and glaciers below.

After 20 minutes I feel as if I'd enjoy just continuing to fly but the Claus lodges appear and Ben brings us down on a gravel land-ing strip. The camp sits on terraces above the Chitina River, cot-tages built of rough-hewn logs, stripped of their bark, basking golden in the glow of the late evening (nearly 10 o'clock!) sun. Paul and Donna Claus named this place Ultima Thule after the mythical northernmost region of the habitable world. But on this trip Ultima Thule is only a rest stop on the way to a territory far-ther north and even less inhabited. Spring's brilliant climax.

That night, around the long table in the main lodge where family and guests gather for dinner, we discuss our plan to explore ANWR, the 19-million-acre Arctic National Wildlife Refuge that lies in the remotest northeast corner of Alaska, some 800 miles to the north. ANWR (pronounced ann-wahr) is bordered on the north by the Arctic Ocean and on the east by Canada's Yukon Territory. On the map Paul traces a route across the east-west arc of the Brooks Mountain Range, a cold and jagged-peaked continental divide that bisects the refuge, north of which all rivers flow into the Arctic Ocean.

I notice that Paul has dressed for dinner, which means he has taken off his baseball cap and the aviator shades that I can tell press close to his ruddy face. He has a boyish look, an impish grin, but his blue eyes are shades too cool to really mistake him for the actor Robert Morse, of whom Paul first reminds me. Paul is a compact package with heavy-duty hands. Donna, robust and always bustling, has a round face that the sun has made just as rubi-cund as Paul's. Both have been divers, commercial fishermen, skiers, and climbers. Paul all but peaked Everest. The St. Elias Wilderness is the Clauses' backyard. It is where they live with their three children (Logan, Jay, and Elie, and Elie's 50 sled dogs) and where their guests climb, raft, and ski. At 17, Elie has already won two junior Iditarods and next year will be the youngest participant ever in the 1,000-mile race from Anchorage to Nome. These are Alaskan kids: home-schooled junior bush pilots, eager for adventure. Jay is spending the spring and summer fishing the salmon run in Bristol Bay. Ten-year-old Logan guns a four-wheeler around the cabins, bare feet to the metal, slinging mud from the big tires, careering with the ease of any kid on a Razor scooter.

Except for Logan, they all have their pilot's licenses, a necessity in this backcountry, a territory made for light and maneuverable high-winged planes with fat tundra tires, floats, or skis: Supercubs, Beavers, and Otters that can glide through narrow mountain passes and land on snow, ice, or gravel, but also spare, light planes like the one I flew in with Ben, one of Paul's young interns, a one-up-front-one-behind plane no wider than a Harley that bears a resemblance to a giant mantis. But seeing Paul's red and yellow Otter on the gravel runway made me realize why the Smiths told me the plane enjoys the admiration of many bush pilots in Alaska.

As we pore over the map I realize that Paul and Donna have flown everywhere in Alaska but ANWR. They're looking forward to the trip as much as I am since they plan to make it a little vacation and scope out the place with the idea of offering it to their guests. Since we'll all be exploring and discovering at once, a decision is made: In keeping with the ideals of Ultima Thule—where, as the Clauses' website puts it, "No itineraries exist. Every day is unique"—our plan is to have no plan at all, but to head north into the refuge, land where we think it looks good to land, camp, hike, raft, look for caribou herds, and move on. I think this sounds perfect. Paul's only disappointment is that the highest mountains are barely 9,000 feet, not much in the way of a challenge compared with the peaks in his backyard.

"We'll see," he says, "what we can find."

The person responsible for most of the names on the map of the Brooks Mountain Range, and who probably climbed and peaked more of them than anyone, was the extraordinary Robert Marshall. Before my trip I'd known of Montana's Bob Marshall

Wilderness, but I'd assumed that Marshall was, like Grinnell, Roosevelt, and Harriman, one of those wealthy dilettante explorers who, as Mark Twain said of Theodore Roosevelt, go "hunting wild animals heroically in Africa, with the safeguard and advertising equipment of a park of artillery and a brass band." It was a terribly stupid assumption.

In preparation for Alaska I'd picked up a couple of books in addition to maps and guides. One was Jack London's *Tales of the North*. After all, I'd be traveling the same territory in which White Fang first met the White men's powder-puff dogs. London wrote that "When a man journeys into a far country, he must be prepared to forget many of the things he has learned, and to acquire such customs as are inherent with existence in a new land." Taoistlike advice but for the implied "or else—" I feel coming at the end of it.

Two other books, *Alaska Wilderness* and *Arctic Village,* were written by Robert Marshall. For some reason I didn't connect the Bob Marshall of the wilderness with the author of the books. I thought that not exploring Marshall's wilderness in Montana was a missed opportunity until I found that Marshall's favorite wilderness was the Brooks Range, where I'm headed. He explored the Brooks in the 1920s and 1930s without fanfare, and he wrote about its places and people with scientific accuracy and an ingenuity untouched by the Jazz Age machismo of his contemporaries London, Grey, and Hemingway.

Marshall was a prodigy. Born January 2, 1901, he grew up in New York City, summered in the Adirondacks (where by 1925 he and his brother George had climbed all 46 peaks above 4,000 feet and on one July day Marshall alone summited 14 peaks, ascending a total of 13,600 feet in 19 hours). He attended Syracuse, then Harvard, and received a PhD in plant physiology from Johns Hopkins. He then went out to explore and map much of the Alaskan Brooks Range, serve as director of the U.S. Office of Indian Affairs and chief of the U.S. Forest Service Division of Recreation and Lands, and write some 100 papers and two books on Alaska—one of which, *Arctic Village,* became a best seller. He fought to have wilderness preserved as public land, and he founded and funded the Wilderness Society. By all accounts he was an avid reader as comfortable discussing geology and plant physiology as he was discussing Russian literature, atomic physics, classical music, Civil War history, or, up in Alaska, the best techniques in hiking or mushing. He was a devoted socialist and a good dancer. As head of Recreation and Lands he worked to find ways to make wilderness areas available to those who couldn't afford to get to them on their own. He was a marvel in the mountains, reportedly going up steep peaks at a fast trot. How, at age 38, his heart gave out on a train from Washington, D.C., to New York, no one knows. Death didn't deter his work from continuing. One-third of his estate went to the Wilderness Society. The rest went to funds to support labor and social justice causes. In 1940, nearly a million acres of Montana forest were set aside as a wilderness reserve and named in his honor.

Like Roosevelt, Marshall was born to wealth. But it was not WASP wealth. The Marshalls were German-Jewish immigrants. Marshall's father, Louis Marshall, was a lawyer who devoted his own prodigious energies to social justice litigation. In the early years of the 20th century he argued and won cases involving voting rights for Blacks and challenged the Ku Klux Klan over lynchings in the South and Harvard University's quota on Jewish students. He fought a withering battle with Henry Ford, whose *Dearborn Independent* newspaper served as Ford's vehicle for advocating virulent racism and anti-Semitism. Marshall, in the end, forced Ford to publicly apologize to the country's Jews in a statement that Marshall himself composed.

It was a tough family in which to excel. So Marshall set his own course, making his interests in science, literature, politics, labor, and social justice serve his fascination for "blank spaces on maps." In the country at the time there was no greater blank space than Arctic Alaska. And it is thanks in great part to Marshall that it remains blank as I get ready to head up.

Most appealing about Marshall's books is that the writing appears to give him as much pleasure as the exploring. His writing has the virtues of great stereo reproduction. He gives the Arctic the presence of a very full soundstage, but one on which none of the subtleties of color and tone is lost. This, for instance, in *Alaska Wilderness*:

It was full starlight when we started, heading straight on the course toward Polaris. After half an hour the black sky began to turn gray, and the unbelievable arctic brightness of the stars slowly faded. The gray became faintly blue, and then a single snow peak in the northwest showed a tip of pink. So gradually that you could hardly notice it advancing, the pink spread from peak to peak until all summits to the north and west were colored. The pink kept creeping down the slopes, changing imperceptibly in color until all at once you noticed that it had vanished, and that the mountainsides were bathed in a golden spray—craggy peaks, snowfields, dark spruce, timber, everything. Then suddenly, at high noon, after journeying a whole morning in the shadow, there was a wide bend in the river, and we drove out into the sunlight.

He never forgets to give good practical details. Marshall answered for me a very mundane but persistent question, which was "What did cold-weather explorers wear before synthetic fabrics and Gore-Tex provided expedition-grade warmth and water resistance?"

I wore a suit of medium-heavy wool underwear, an ordinary flannel shirt, a pair of light wool pants, a pair of cotton overalls (to break the wind and keep the snow from sticking to the wool), a sleeveless sweater, one of those closely woven, green

Filson cruising jackets, and a light cloth parka over everything. The parka had no buttons, could be slipped over the head, came down a little below the knees, and had no vent to let the wind get through. On one or two fairly cold days—it was never more than 31 below—I wore a caribou-fur parka while riding the back end of the sled. For my head I used a muskrat cap with pieces covering my ears and the sides of my face. There was a hood on the parka which came way over in front, and was a great protection from the wind. On my feet I wore three pairs of wool socks, innersoles, and moose-skin moccasins with eight-inch tops. My hands were protected by anything from wool mittens inside and wool-lined leather mittens outside, to just canvas gloves, depending on how cold it was.

I have a suit of polypropylene underwear, a Capilene turtleneck, a pair of Polarfleece-lined jeans, a polyester fleece jacket, a down vest, and a Recron Fibrefill parka with a water-resistant microfiber coating. It being spring I'm hoping it will not get to 31 below. But still, I've got a Gore-Tex shell, nylon undersocks, polypro hiking socks, and Gore-Tex-coated boots and gloves.

I'd also packed overpants, shirt, and hood all made of a fine mesh in order to appear as much as possible like a screen door to the more than two dozen species of Arctic mosquito, their spring swarms variously and harrowingly described in the literature as "voracious armies," or "oppressive hordes" that can "emerge explosively" and "drive an animal or human insane." These even drove Marshall to distraction.

We spend today gathering provisions, which out here requires flying from one place to another. Up along the river we check in on the Clauses' fishwheel. It is an antique device for catching fish in the current of a fast-moving river. It requires neither fuel nor electricity and has been used out here for salmon since at least the late 1800s and has not changed much since. In its most basic arrangement a fishwheel consists of two or three wire cages that revolve, Ferris wheel–like, around a center post that is suspended out into the river on floating pontoons. The river current turns the wheel's cages, which scoop up fish swimming upstream. As the cage rises out of the water the fish slide off down a chute and into a wood box or holding pen. Simple, cheap, effective, and even portable, Alaskans living along salmon-run rivers usually have their wheels up and running early in June for the first runs of chinook and sockeye salmon.

The Clauses haven't had the time since last season to set up their wheel properly and so there is no catch. It's not for lack of fish. At the fishwheel next to theirs stands a rather large red-faced man with a three-foot-long salmon laid out on his cutting board. Without stopping his fish-gutting, unable to speak for the cigarette pursed in his lips, he lifts one leg and kicks open his cooler to show us three more salmon inside. Finished with the fish, he puts down his knife, takes out his cigarette, and lifts the fish by its gill flap for our consideration. Looking into his cooler and back at us, he says, "Take it. I've got plenty." We proffer polite demurrals but, as we hope, he insists, takes a plastic bag out of his truck, slides the fish in, and hands it over. "Good eating."

To me, this passing of the fish was in its way the marking of a spring rite. It didn't have the mythological bearing of a Tlingit or Haida First Salmon Ceremony but I think it was the best that animistically challenged White men could do. Sadly, the native ceremonies have been dying out as surely as the salmon, whose arrival once marked the end of vanishing winter food supplies. Once, come May, vast schools appeared in the thawing Alaskan coast bays, fish literally springing from the water as they surged up the mouths of the rivers. As the salmon arrived, so did the shorebirds. Bears came out of hibernation to feed on the fish. Bald eagles that had returned north began nesting, feeding both on birds and fish. The first wild berries began to appear and the sweet young shoots and roots of plants could be dug from the thawing ground. It wasn't much of a leap to ascribe all of this activity, the coming of spring, to the salmon.

On rising spring tides these fish, having spent the last few years out in the Pacific, press their way upstream, up the Copper River to the Chitina, for instance, going hundreds of miles if necessary, seeking the waters of their birth. Upon finding them they'll spawn among the gravels of their home shallows. By then they're thin, pallid, frayed-finned, and gape-jawed. This first return journey will most likely be their last. To partake of the first salmon was to partake of the life it brings. In other places and times the same could be said of first bread, first corn, first fruits. Frazer writes that in May the Aztecs would shape bread into the image of the god Vitziliputzli. Worshippers believed "that by consecrating bread their priests could turn it into the very body of their god, so that all who thereupon partook of the consecrated bread entered into a mystic communion with the deity by receiving a portion of his divine substance into themselves." And this, Frazer states, took place before the arrival of the Spanish and Christianity. Or perhaps it was a Christian interpretation of what the Aztecs were doing. It is never easy, when journeying, to see things the way others see them. It may not be possible at all.

So what if you had to try to see things the way the world of nature sees them? This is precisely the concern facing native people of Alaska who imagined a world that was itself aware. In *A Koyukon View of the Northern Forest* (the Koyukon are northern Athabaskans), Richard K. Nelson writes that for the Koyukon, "each animal is far more than what can be seen; it is a personage and a personality, known from its legacy in stories of the Distant Time. It is a figure in the community of beings, once at least partially human, and even now possessed of attributes beyond outsiders' perception."

The salmon ceremony is not the worship of nature or the divine. It is at once far more supernatural and far less abstract. Its purpose is to maintain good relations between the human and natural worlds. The salmon ceremony says, "Thank you for bringing the return of life; what you've brought us we'll share, good journey, and farewell."

That night the fish is served. Paul and Donna say a grace before the meal. They are born-again Christians and so I imagine that they have an even different sense of what life's revival is all about. I'm sure it will make for some interesting conversation on our trip. The fish is fine, like river-run venison.

At 10 o'clock it is still light enough to sit on the porch of my cabin and read Marshall.

> I do not know what may be the supreme exaltation of which a person is capable, but for me it came that moment I crossed the skyline and gazed over into the winter-buried mystery of the Arctic, where great, barren peaks rose into the deep blue of the northern sky, where valleys, devoid even of willows, led far off into unknown canyons.

Here is a thought to dream on.

June 17
19 HOURS, 20 MINUTES OF DAYLIGHT

The Otter, its tanks filled with fuel, its expansive hull filled with gear and food, is ready to go. We climb over the stuffing into our seats, strap in, put on headsets so we can talk to one another, and take off heading northwest in a light mist, following the river valleys and seeking out blue sky amid the surrounding cumulus. Paul finally finds an opening above the cloudcover at 10,000 feet. From this vantage we can see only the rugged peaks of the highest mountains. If you can imagine this cloudcover as the surface of a vast snowfield, 8,000 feet deep, then you know what Alaska looked like 10,000 years ago, when the land all lay buried beneath glacial ice sheets. As we pass the eastern edge of Denali National Park we bear northeast through the mountains of the Alaska Range, where our patch of blue sky expands. The low ceiling beneath us falls away and reveals the changing aspect of the land below—the sprawling green valley of the Yukon River, a 200-mile-wide taiga where lakes seem to cover as much of the land as the dense stands of spruce.

The Yukon and Porcupine rivers meet among braided streams and looping oxbows that twist through a land as cantankerous as the gold rush sourdoughs themselves. They came in search of rich strikes and their hopes would rarely pan out. They came on the heels of the fur hunters, the Russians, the Hudson's Bay Company, all of whom considered the native Athabaskans—whose ancestors migrated across the Bering Straits some 10,000 years ago—as useful but "uncivilized." "Uncivilized," in fact, was the natives' official status once Alaska passed from Russian to U.S. territorial government in 1867. Be that as it may, the Athabaskan languages spread across the Northwest. The language, changing as it went, also formed the basis of the Indian languages in the Southwestern United States. What remains of those connections may be heard in songs and melodies shared by the two distant groups. This was a far-flung culture.

We're about 160 miles northeast of Fairbanks, over Circle, Alaska, on the south bank of the Yukon River. Named Circle because the miners who established the town during the first Alaskan gold rush mistakenly thought it lay on the Arctic Circle, it is actually 75 miles south. The town served as a transfer point for goods headed for the mining camps, and by sub-Arctic turn-of-the-century standards, it flourished. By 1896 the town boasted 700 residents who had in Circle just about all even a civilized sourdough could ask of a place: library, school, hospital, newspaper, Episcopal church, a dozen dance halls, and 28 saloons. When its hewn-log opera house opened, Circle became known as the "Paris of the Yukon," and the town's bohemian cast of characters served as muses and models for the Yukon's muscular poets laureate, Jack London and Robert Service.

When the Klondike gold strikes came, the miners, though reluctant, I'm sure, to miss the next performance of *Les Troyens,* moved on to Nome and all but abandoned Circle. The town now has all of 100 residents, most Athabaskan Gwich'in. Gold was never their purpose in Alaska. The Gwich'in are the "people of the caribou." They call the coastal plain of the Arctic National Wildlife Refuge "the Sacred Place Where Life Begins." From here and from the towns of Arctic Village and Fort Yukon, across Arctic Canada, the Gwich'in still hunt the porcupine caribou, named for the river the herd crosses on its migrations to and from ANWR.

Claus informs me we're a few miles from Fort Yukon and a few miles north of the Arctic Circle. He flies low over the river's tangle of meanders and suddenly drops out of the sky to make a soft landing on a gravel bar. As we set up camp I see my first Arctic terns, swooping over our heads, warning us away from their shallow nests in the gravel and sand. A dowitcher struts in a tidal pool. Purple lupine blooms along the riverbank. Donna picks wild onions while we collect driftwood and build a fire. The light lingers. I've made it to the Arctic spring.

I decide to take off my watch to make the most of these last extended hours of spring sun. We cook and talk. Paul says that tomorrow, with good weather, we'll refuel in Fort Yukon and head into ANWR. We go to sleep with the sun looming through the high haze as if it were dawn.

June 18 to 21
96 HOURS OF DAYLIGHT

First thing today is a fuel stop at Larry's Flying Service in Fort Yukon, where the sign reads:

Fort Yukon
Inside the Arctic Circle
Lat. 66.34 Long. 141.51
Pop. Summer 800 +/−
Pop. Winter 700 +/−

Inside the office an old-timer, who looks like he stays around summer and winter, warns me with Alaskan bluntness that ANWR is not the blushing country that "it's made out to be by those environmentalists. It's bare and mosquito-ridden," and, echoing the Arctic's poet laureate Robert Service, he calls it a "godforsaken place."

"Well," I say, trying to keep things light, "godforsaken is okay with me."

This doesn't lighten things up at all. He nods as if I've just confirmed his sorry opinion of me, slaps his knees, stands, sighs, and shuffles out the door. I might have told him that I've traveled often in the north and have no illusions about what I'll find. But this would not be true since I have tremendous expectations, visions of vastness, brilliance, and timelessness. These hopes are not dashed as we fly north and get our first look at the Arctic.

Paul navigates along the river valleys. Heading up the Sheenjek River, the dense groves of black spruce that we'd been seeing thin out and then disappear altogether. Few trees can survive the winter cold and dark this far north. Those hardy specimens that make it grow so slowly that after a hundred years they may be no more than several inches thick. When the trees vanish there is only taiga and potholes in which the water is a turbid lava-lamp green from the growth of spring algae.

The southern foothills appear like an ashen Sahara, where bare massive dunes of gray slag extend for miles. The only signs of life are sedges growing along the runnels of meltwater. Ahead of us the Brooks Range (named after geologist Alfred Hulse Brooks) rises to a 9,000-foot knife-edged crest, cold as cast iron.

If, as Service suggests, God created the Arctic and forsook it, His exasperation was nothing compared with that of the geologists who've tried to decipher this landscape's corrupted stratigraphies. Crumpled, twisted, shifted, uplifted, infused by lava flows, deformed, eroded, faulted, folded, flooded, obducted, metamorphosed, crystallized, recrystallized, and glaciated, the formations of the Brooks Range include rocks as old as 570 million years and as recent as 66 million years, although not necessarily in any particular order.

We tack east and cross the mountains' continental divide beyond which all rivers flow to the Arctic Ocean. I'm feeling pretty giddy from the whirl of scenery on our low and swooping flight path. And yet there's more. The Kongukut, the refuge's easternmost river, appears below and we dutifully follow its gleaming turquoise flow northward. The valley widens. We fly even lower. The river twines its way among sandbars and stalled ice floes. And Donna calls, "Caribou!"

We see them along the banks of the river, grazing singly or in small groups. Some stand on ice floes. We're right on the Porcupine herd's main route. The animals we see are on the final leg of a three-month-long spring migration that began in April when they left their Canadian wintering grounds. This long bend of the Kongukut is known as Caribou Pass. The broad Arctic Ocean coastal plain that is their final destination lies some 25 miles farther north. There, for a day or two, the herd may number in the tens of thousands before they disperse to feed. If we're lucky, we'll meet up with them. The females needed to reach the plain early to give birth, so the animals we see here are most likely males straggling behind.

A plant or animal can be in only one place at a time. As self-evident as that seems, we too often assume that the world of a living thing is only where we find it. The truth is that, just as humans live one place, work in another, and travel through a variety of places going from home to work, most animals and plants do the same. The area over which a living thing lives, feeds, and breeds is called its range. Plants that produce pollen need their pollen to spread over a range in order to produce new plants. A 400-pound grizzly bear ranges over some 7,000 acres for its food. A migrating salmon ranges from the ocean to the headwaters of freshwater streams. Monarch butterflies range from Mexico to Maine. Those aggressively protective Arctic terns spend three months in the Arctic nesting and brooding then take off and three months later land 12,000 miles away in the Antarctic, where they winter—again in 24 hours of daily sunlight; in all they range over 24,000 miles of land, air, and water. Each spring young eels that hatch in the Sargasso Sea in the Atlantic Ocean find their way a few thousand miles to rivers along the Atlantic coasts of the United States and Europe, where they move upstream to mature. Everywhere, the natural world is made up of overlapping ranges—plant and animal and human. And in spring they're all seeing action, every court's in play.

Usually, when human ranges cross those of animals and plants, the latter suffer. But not always. The Arctic Porcupine caribou herd, some 120,000 animals, ranges across an area the size of Wyoming. That this is also the historical range of the Gwich'in people is no accident, for the Gwich'in have lived for at least 8,000 years along the Arctic Circle in communities from northwesternmost Canada west across the southern foothills of the Brooks Range. For all of that time they've hunted, fished, and done what farming and gathering they could along the Yukon Flats, but their lives depended on the caribou's annual appearance. During migration season they eat caribou three or four times a day. Boots and clothes are made of their skins. Bones serve as tools. The Gwich'in's 15 villages all lie along the caribou migration route. Their myths speak of human and caribou not only sharing one world, but that every caribou has a bit of the human heart in him, and every human has a bit of caribou heart. In this way the ranges of the Gwich'in and the caribou overlap spiritually as well as physically.

It has been a relationship that has lasted through difficult times. No time, however, has been more trying than the present. When, in 1971, tribes in western Alaska agreed to allow, in exchange for money and secure lands, oil development and the Alaska pipeline on their territory, the eastern Gwich'in refused, deciding instead to continue fighting for original land claims while living the subsistence existence they'd negotiated successfully for thousands of years. While they knew that shunning the state and the oil industry was not politically or economically expedient, they didn't count on the possibility that oil drilling might come to ANWR's coastal plain, where damage to the caribou herd would threaten the Gwich'in's existence. (The Alaska state house, in their appeal to the U.S. Congress to bring oil drilling to ANWR, removed a reference to the relationship of the Gwich'in and the caribou. Politics is less forgiving than wilderness.)

Bob Marshall wrote, "There is just one hope of repulsing the tyrannical ambition of civilization to conquer every niche on the whole earth. That hope is the organization of spirited people who will fight for the freedom of the wilderness." The Gwich'in have proven themselves "spirited people." But the niche tyrants are not easily deterred.

And what about us visitors? Some of ANWR may be as "godforsaken" as my old-timer told me. (He also warned that "naked, the mosquitoes would suck you dry in a couple of hours." I intend to keep my clothes on.) But ANWR wasn't preserved as a place for us to feel at home. As far as the caribou, wolves, bears, and birds are concerned, the more cussed the place is, the better. Yet the truth is that in spring, an imaginative, well-prepared human can, for several weeks of the year, make himself quite at home in ANWR, a place as unlike home as any place one is ever likely to travel.

One small step for Man . . . , I'm thinking as I step down the plane's ladder onto solid ground. Paul has made yet another sweet landing, this time on a narrow gravel strip on the bank of the

Hulahula River (named such by homesick Hawaiian whalers). Once the prop stops it seems to take a few moments for the valley to piece together the shattered silence and for the drone of the plane to fade from my own head. My first footstep on the gravel has the effect of crinkling a candy wrapper in a concert hall. So I stand still and take in a view so vast—so far I'd seen it only from the air—I feel as if I've been digitized into the landscape. Without trees to obstruct the view the scene is a study in vanishing points: "Winding River, Distant Peaks." On both riverbanks the bases of the mountains swell into heathery slopes. Above these rise rugged bare climes of crumbling shale. Awash in light the faces of the most remote mountains, perhaps 50 miles away, loom strangely two-dimensional.

With my watch packed away I have no idea what time it is. The sun is bright. But only two days away from the solstice the question is which sun it is—the noonday sun in the south or the midnight sun in the north? How long had we been flying? Six hours? Ten hours? We unpack the plane and set up camp. Donna and Paul have a spacious orange vented cook tent that, with the red plane nearby, makes for a very lively campground. With the temperature in the 60s I decide not to sleep in a tent at all and lay my sleeping bag down close to the river. My theory has always been that you go out to be out so why be out and indoors? Mosquitoes could change my mind but an upstream breeze is keeping them at bay and blowing a fine dust of gravel and sand. The air is dry. The north slope of the Brooks Range gets as little precipitation as the Arizona desert.

Donna must know what time it is because she decides to cook dinner. Our conversation turns on Paul and Donna having spotted from the air a few groups of rafters out on the Kongukut. Paul is disconcerted.

"In the Wrangells I still land in places no one's ever been before," he says.

I think to say that a dozen rafters up here hardly constitutes a crowd, but then I live in Brooklyn. Besides, Alaskans still measure their freedom by how far away they can get from the closest human. This question seemed on the mind of every adventurer about whom I'd ever read, Arctic or otherwise. Perhaps this is how we define freedom and solitude. Or does it just enhance our sense of proprietorship? That is, if there is no one else around do we imagine that all we survey is ours? Sometimes this "being a hundred or a thousand miles from the nearest human being" is the equivalent of having the biggest house, pool, or car. In any case, all of this distorts the idea of a wildlife refuge. I know people in New York City for whom a few square feet around a tree in Central Park is refuge enough—a place where they can read and not be hit by a soccer ball. But the question is not how much wilderness *we* need but how much wilderness we're willing to provide to fulfill the needs of the wilderness. I figure ANWR belongs to all of us and none of us. Whoever else is here has accepted the conditions of the place. They've been deposited here, as we have, by plane and left to hike, raft, camp, until they're picked up and whisked away. They leave behind no cabins, summer homes, or tire tracks. Unlike, I suggest to Paul and Donna, the oil companies who persist in wanting to drill for oil in the refuge. To begin such a conversation in Alaska is to step out on increasingly thin ice. For many Alaskans, drilling for oil is no less natural a pursuit than fishing for salmon.

Paul is diplomatic. "It's different now than it was," he says. Years ago, certainly, the oil companies left a maze of roads across the tundra. (We've seen them from the air.) But now "the oil companies know how to work without damaging the land." Besides, he points out, this place is huge and they'll only use a percent of a percent of it.

That is true enough. So what if the Arctic Refuge was only 29,000 and not 30,000 square miles?

Last spring, in what appears to be a congressional spring rite, an energy bill loomed that included yet again a provision to open

ANWR to oil exploration. During the summer, party lines broke over the provision and by fall the bill's sponsors pulled it. The refuge would remain untouched and the coastal plain known as the 1002 (ten-oh-two) Area safe for another year. The decision did not sit well with most Alaskans. They do have a stake in the issue. Oil companies provide jobs and their revenues are the reason Alaskans pay no state taxes. I'd rather they say that straight out than persist in peddling ANWR oil as the cure-all for the nation's energy ills. The truth, if one believes the reports of the Department of Energy, the U.S. Geological Survey, the U.S. Fish and Wildlife Service, and the National Academy of Sciences, is that drilling in ANWR is a bad bet all around; that even once the wells are pumping to capacity, perhaps within 20 years of their opening, the country's increasing oil demands will be such that ANWR's puny output will barely supply 1 percent of our daily oil consumption, an amount we could easily make up by conservation.

And despite claims for new, improved, and environmentally sound methods of exploration and drilling, the U.S. Fish and Wildlife Service concludes that "development in the 1002 Area could likely require a large number of small production sites spread across the refuge landscape, connected by an infrastructure of roads, pipelines, power plants, processing facilities, loading docks, dormitories, airstrips, gravel pits, utility lines, and landfills." All of this in an area "critically important to the ecological integrity of the whole Arctic Refuge," that "provides for greater plant and animal diversity than in any other similar sized land area on Alaska's North Slope." This is the place, after all, where Arctic life is born.

The Alaskan powers-that-be and their allies swear to battle on in what they admit is a mostly a symbolic fight.

"A thing is right," wrote biologist Aldo Leopold, "when it tends to preserve the integrity, stability, and beauty of the biotic community. It is wrong when it tends otherwise."

The argument over ANWR is no longer about "what's good" or "what's right," it is about "what's mine."

"Wilderness is fine," they're saying, "but don't ever mistake it for the caribou's, wolf's, bear's, elk's, Gwich'in's, or rafter's wilderness." Even in 1937 Marshall could see this coming:

> Over most of the United States, the prevalent attitude has been that the greater the development of natural resources, the greater the public welfare. Allowing a resource to remain as it had remained for millions of years was considered anti-social . . . Many invaluable resources have been depleted disastrously. In Alaska the dominant development policies of the United States should be balanced by a policy of preservation.

Further:

> . . . the country of northern Alaska is inhabited chiefly by native populations which would be much happier, if United States experience is any criterion, without either roads or industries. Therefore, I would like to recommend that all of Alaska north of the Yukon River, with the exception of a small area immediately adjacent to Nome, should be zoned as a region where the federal government will contribute no funds for road building and permit no leases for industrial development.

This was the recommendation of the man who was then in charge! But since the time when Marshall was head of Forest Recreation, the climate has changed, in more ways than one.

Whether it is the next morning or not, I can't say for certain, but we awake, eat breakfast, and decide to hike. Paul has seen Dall sheep grazing on high ground above the river and thinks that if we climb and come around the back of the ridge we might get close to them.

Surprisingly, hiking on the low slopes is hard going. What appears to be solid ground is not. Rather, it is made up of small sedge tussocks that are like tiny haystacks with the spaces between them often filled with water. This water, from rain or melting snow, has nowhere to go since the ground only several inches below is permanently frozen. In spring, however, the life of the tundra takes place in those few unfrozen inches above the permafrost.

Higher up we reach jagged and brittle slate. Every footstep threatens a small avalanche of crumbling oily rock. The temperature is near 70 and I foolishly forgot to bring water, thinking there'd be plenty of it on the mountain, and now I feel weak in the knees. For my sake we detour to a stream of meltwater a little lower on the slope and drink. With the light never changing I have no idea how long we've been out. I imagine we've climbed 1,500 feet above the valley. Looking down at the river below, our plane and tents are red and orange Cracker Jack miniatures. We've got maybe another thousand feet to go to reach the stack of rock on the hill crest that points directly up to the sun. I look up the rock face and see the sun ringed by a transfixing rainbow—light refracted through the prisms of ice crystals that float invisibly in the Arctic air.

We change direction and traverse the face of the slope, and as we come around Paul somehow spots the sheep. I finally see them through my binoculars, three of them ensconced and still in a cliffside nave, two facing right, one facing left, like high reliefs on a temple frieze. Paul has warned us that out here the animals will readily hear or smell our presence. Unless we take time to approach, they'll hie off. By time, he explains, he doesn't mean minutes. Out hunting, he says, he has spent hours approaching an animal. One time, he recalls, he spotted a caribou lying on the ground by a large rock. Over the course of two hours he used all his tracker's wiles to sneak up on it, zigzagging to stay upwind, lying low, moving step by step until he was close enough to see that the animal had died some time ago.

With no noticeable change of light I have no sense of how long we've been hiking. At the spring equinox on March 20, the Sun appears to be crossing Earth's equator. In the temperate zones day and night are of equal length. At the North Pole it is the beginning of six months of daylight. Since Earth is tilted, those 23 degrees relative to the Sun and the Arctic Circle lies 23 latitude degrees south of the Pole, it's not until June 6 when the Arctic sees sunlight 24 hours a day. It will last for 29 days. Instead of the Sun seeming to move from east to west horizons it appears to circle the sky. If you watch for it, you can recognize the movement by the barest shifting of shadows, especially along the sharp cusps of the high ridges.

When we get back it turns out that we've been hiking for eight hours. Much to Paul's distress another plane has landed not more than a few hundred yards away from us and deposited a group of campers and their kayaks. He comes into the cook tent visibly tense, makes himself a peanut butter sandwich, and, while he eats, pores over the topographic map. Practiced in landing on peaks half again as high as the highest in ANWR, I have the feeling he is looking for something to challenge his artistry as well as to bring us to a place where we're certain not to see a soul.

While my own sense of time seems to be slowing down, Paul continues to move as if we haven't got all day, which we do. He finishes his sandwich, looks over at me, and says, "You ready?" No sooner said than we're back in the Otter and alone in the sky. Paul is after some fun. We see a long line of sheep heading downslope to graze. He spots a couple of grizzly bears below; one, to his impish delight, exploring the thickets near the river not far from another rafting camp.

"They can be that close," he says, "and you'll never see them."

Like a butterfly from flower to flower, Paul wings the red plane from peak to peak until he decides to alight in the snow-covered glacial cirque of Mount Michelson, at 8,500 feet the next-highest point in ANWR. What's left of the glacier on Mount Michelson pours down from an exposed crest of chiseled granite. Nearly all of Alaska's glaciers are in retreat, the signature of a warming Arctic world, melting away and leaving bare rock behind.

Paul scans the glaring field of snow below while he works out a calculus of variables that, he tells me, include not only wind speed and direction—often erratic up close to the mountains—but also the outside temperature, which tells him the condition of the snow. He dips right down into the half hollow of the mountain, banks, veers off from the wall of rock, lifts the nose, and comes around again. His left hand on the wheel, his right hand quickly trims the flaps and brings the skis into position below the tires. Coming out of the sharp bank he drops onto the snow. We slide no more than 25 yards before coming to a gentle stop facing downslope. Paul proudly announces ours as probably the first-ever landing here of a fixed-wing aircraft.

We step down from the plane, stand within the bowl of the glacier in a kind of joyful daze. Paul and Donna hold hands like young lovers in spring and stroll to the base of the jutting granite escarpment, where Paul halloos up into the rock to hear his echo.

I decide to walk up to the snow-blanketed ridgeline. Snow at this altitude is so reflective that the sunlight barely warms the surface. As I walk across the glaring white plain toward the white crest above I realize that without shadows to distinguish ground from rising slope there is no perspective, no way to calculate the distance. As I walk, however, the sun moves and the bank above appears to change shape, at first softening, then taking on the curl of a breaking wave. Near the top I recognize the crevasse where the glacier ends and it is only deep snow. It seems to me that I cover a long distance in a short time but with time and distance so variable I couldn't say for certain how far and how long. Over the last days both have seemed to matter less. When I turn to head back to the plane, Paul and Donna appear as tiny figures on the snow below the high jagged arête. We fill our water jugs with glacier snow and take off.

We can't determine the name of the next mountain we land on—it may have none—but it is nearly as high as Mount Chamberlain, the highest peak in ANWR and in our view just to the northeast. Rather than being scooped out, the top of the mountain is a dome of snow and Paul puts us right in the center. As we walk the perimeter of this sky-scraping island there is nothing to obstruct (or way to describe) a view that extends at least a hundred miles in every direction: across the northern foothills of the Brooks to the fogbound Arctic Ocean shore; across the descending peaks east to Canada; and across the craggy, snow-crested mountains south and west. Call it, as the 18th-century writers might, "an infinite," perhaps "ethereal vast." Nowadays, "a View-Master vast" or "an IMAX vast." But the seeing is only a part of it—the part that takes my breath away for sure, but even more compelling is the dynamic juxtaposition of ice peaks, flowing waters, spreading valleys, and cold distant ocean that engages the imagination.

I ask myself how I feel up here and once I get past imagining I feel what I'm supposed to feel, I realize I feel neither humbled nor insignificant; just the opposite, I find my senses trying to meet the challenge they're being presented. After all these months chasing spring here to its apogee I know that however still and immutable this landscape may seem, it is so full of energy and transformations that it is practically jumping out of its frame. To the right ears the sound of photosynthesis would be deafening. Perhaps I'm actually seeing time and space. I try to recall whether physicists found the universe finite and unbounded or infinite and bounded and wondering which this is. Maybe I'm feeling high and mighty from the altitude, but I conclude that the universe of spring is neither Euclidean nor Newtonian.

"The concept of space as something existing objectively and independent of things belongs to pre-scientific thought, but not so the idea of existence of an infinite number of spaces in motion relative to each other." This comes close. Einstein and I on the mountain, up here where in spring "us physicists believe the separation between past, present, and future is only an illusion . . ."

We fly: down from the summits of the Brooks Range across the rolling foothills and out onto the sprawling tundra of the coastal plain. Here, the many streams that rush off the mountains become mired in the sand and gravel they've pushed downstream ahead of them. They flow where they can amid dense thickets of dwarf willow. Paul spots some caribou and lands us on the lower Kongukut, about a half mile away from the herd. He gauges the direction of the wind to keep the caribou from smelling us and we begin our hike across gravel beds and through the streams that separate them. Beyond the braided channels that we're slogging through lies a low rise, a meadow of greening tundra. If we can reach the last hedgerow we'll be behind a blind barely a hundred yards from the herd. We stay low and say nothing. We're close enough now that we're practically crawling through the dense thickets. When we're as far as we can go we realize the only place to stand will be in mud and water, but we've come this far and the view through the blind is like peering into a great cathedral through the crack between the doors. As it turns out, we're just in time to see the congregation filing in.

From every low slope we can see, we see them coming. Herds of hundreds joining the aggregation of thousands. They saunter in loose formation, large deer on gangly legs, many with fluff-tailed young—no more than a couple weeks old—scampering alongside, some trying to grab a teat and suckle on the run. Once and again an adult raises its head and pricks up its ears, seemingly aware of our presence, but then moves on, perhaps feeling safe among such numbers. Hunkered down below the willows, wet and cramped, we watch their march for hours.

And for hours under a vast blue sky they continue to arrive, pass by our blind, and move toward the sea, gradually becoming silhouettes on the distant flat plain, the place where life is born. Their spring journeys began about the same time as mine, and here is where both our journeys end. The synchrony in this place in this time of never-ending sun fills me with inexpressible joy. My eyes burn from glare, wind, and tears.

We've been lucky to see them. Although most years half to three-quarters of the herd gathers near the Kongukut, exactly where they'll aggregate and when, when they'll move and to where, may change according to the herd's own demands. Efficient travelers, they won't remain gathered here for long. In a day, maybe two, they'll disperse along the coastal plain to graze on the shrubs and the tundra plants that have reached their peak of growth. Then they begin their long return migration south. Neither we nor they have yet been assaulted by the mosquitoes. When they are they'll go down close to the ocean for relief. This coastal plain makes up only 10 percent of ANWR and perhaps less than that of the Porcupine herd's range. But the herd's survival depends upon being able to freely graze these lands. How soon will they face the same fate as so many other species—forced to migrate through landscapes that humans have altered, constrained into ever smaller reserves until faced with extinction?

My last day in ANWR we take inflatable kayaks down the Canning River, the western border of the refuge. The water is swift, the rapids gentle, and only the upstream winds make the paddle challenging. The landscape, as everywhere, is daunting. But best of all I meet up with the birds whose journeys also end here. We drift past thousands of terns and gulls nesting along the gravel bars in the river, sandpipers nestled side by side, their white heads lining the bars as if it were a parade route.

Time slows to a crawl or doesn't seem to move at all. We camp that night on the Canning and after dinner, after sleeping some, I awaken—perhaps in the middle of the night if it was night—and climb the low slope above the river. On my way I pass grizzly bear tracks. The tundra is greener than it was a day ago. The tussocks seem to have suddenly burst into bloom. There's a pleasant citrusy scent in the air. I sit—I now hardly notice the silence—and stare at the ground around me. A half dozen and more species crowd onto each small tussock, and each is in flower. Lupine, forget-me-not, heather, anemone, campion, saxifrage, dryas, and poppy—a garden of miniature wildflowers that sprouts from a pincushion. I swear to myself that at my feet lay the most beautiful few square yards of ground I have ever seen. Unlike Robert Marshall, I feel more drawn to the depths of small worlds. Within them lie all the great complexities of vaster places. I hike higher. And I swear the flowers are blooming right before my eyes. And I think of the millions of acres of tundra meadow, and that across the Arctic all of them on this last day of spring are in just as intense a bloom. Suddenly these tiny plants loom as large as the mountains on which they grow.

I think the difficulty that I face having reached the end of my journey is that I've begun to suffer a kind of Stockholm syndrome—so long in the thrall of spring that I've begun to identify my existence with that of the season's, imagining that all of spring's transformations, enticements, multivarious sensual and fragile beauties (for which I've been an obsequious sucker) have all been proffered for my benefit. Each day I watched light and saw it transmute into structures that, breath by minuscule breath, synthesize the earth's food and by their infinitesimal exhalations create its atmosphere. I've fallen in love with the spring of my own being. The easy part of this love is that it demands nothing of me. The hard part is that although it controls my life it has no interest in me at all.

Our own bodies are no different—functioning assemblies of cells and fluids that have no more interest in the whole that they're a part of than the engine of the Chevy cares who is driving, than a heart cares whose heart it is. Love, understand, or experience nature. Malign, misapprehend, or ignore nature. What you will, whatever we do with our sources of life, refuge, or imagination, we do for our own or the world's good or peril. If you're seeking a god or an intelligence behind it whom you can imagine designed red flowers to go with green leaves, who fashioned time and space for your needs or their pleasures, or who hears your prayers, hosannas, or pleas, it is not nature.

Poets write on spring because this season, like no other, lays bare all the makings of epiphanies—integrity, synchrony, and energy. The poets of China and Japan seem to get it right most often. Or at least the bad stuff hasn't survived the centuries. After all, wholeness is the Buddha's milieu, if one can say that, and the Asian poets expressed that wholeness with economy and precision in poems with such integrity that a single changed syllable would, like an altered strand of DNA, alter it all. The 17th-century Japanese poet Matsuo Bashō could reduce the wholeness of a world into a tension-filled, seventeen-syllable microcosm:

> *Spring:*
> *A hill without a name*
> *Veiled in morning mist.*

This hill, this land, what will happen here? No one seems too sure. The Arctic is warming far faster than any place on Earth. In the last 30 years winters have warmed by 4 to 6 degrees Fahrenheit. Where the temperature of the ground averages near freezing, the thawing can have some extraordinary effects. Beneath the tussocks and few inches of soil in the Arctic lies permafrost, a frozen layer of ice and rock that can be a few to hundreds of feet thick. Since anything built in the Arctic stands on a foundation of permafrost, the melting will begin to wear away at structures, from roads and homes to supports for oil pipelines.

Thousands of years of organic material, ancient carbon (perhaps 14 percent of the earth's total carbon) lies within the permafrost, stored like food in your freezer. Should the permafrost melt and bacteria begin to break down that material, more carbon will be released into the atmosphere and further increase the rate of warming. When? By how much?

If the permafrost begins melting, the water that usually drains off the tundra will pool and, come spring, turn the tundra to soft mud. The change may be exacerbated by increased rainfall in the Arctic. Studies in Norway have already shown that reindeer forced to traipse through mud for food end up eating much less than they need. Warming and softer ground will also introduce new plants, insects, and even birds that before might not have survived the brief warm season. Warming has already caused the sea ice to retreat and the coastal land to erode and threaten the survival of villages along the shore. Arctic sea ice is more than three feet thinner than it was 30 years ago. With sea ice vanishing, polar bears find fewer places to hunt and may vanish altogether from the Arctic.

Change happens. But there are two aspects of these recent changes that make dealing with them imperative:

First, in all of the past periods of climate change the human population that had to cope with the consequences was either nonexistant or relatively few. We've now got to deal with pervasive change on a crowded planet populated by people who have expectations beyond basic survival.

Second, physicists know that small disturbances in large systems can provoke sudden chaotic responses that will abruptly accelerate change. In ecosystems the shift from gradual to sudden change may make the process irreversible: an epidemic might emerge, fisheries might collapse, climate shifts might prove catastrophic.

Spring might become the season of uncertainty—of storms, floods, thaw, or drought—and not of longing, seeking, finding, greening, growing, awakening, of poetry, kinesis, light streams, and ancient genes that inspire cascades of molecular change.

I lie in my sleeping bag, the crystal blue river runs close by, terns and gulls wing low and skirt the wind. Above, the continuous sun, the yellow of powdered sulfur, as much light as aura, feels like a prize won. What is best about this place I've come to is that it is not meant for you or me at all. For us, it is a place where, for the moment, past, present, and future are all one and inexorable, where the senses and the spirit find refuge.

But for how long?

Timelessness isn't what it used to be.

What will your and my children's and grandchildren's springs be like? Will the poetry of the season still make sense to them? Will spring remain as full of energy and omens and just as radiant and manifest? Will they be able to head out in spring to recover their hearts and find places like this to lie out alone in the open and ask themselves why things—why they themselves—are where they are?

As my grandmother would have put it, "I've been and gone," not "come and gone," but actually established my being somewhere before going. It won't make leaving this bright place, this luminous season, any easier, but come February when snowmelt begins, when the soil moistens, when the first warm rains fall, I know I'll be thinking salamander thoughts, ready for the vernal rush, more aware and more appreciative than ever of each new minute of light that marks the rising arc of the sun.

See spring for yourself. See spring now, because it is changing.

All things are changing beneath the sun. See it now.

Sources

Thanks to all those researchers who were willing to share their work and time with me: Neil Tyson; Michael Klemens; Jim Malusa; Mark D. Schwartz; William Schlesinger and his colleagues at Duke University's Nicholas School for the Environment; Mark Williams, Tyler Erickson, and their altitude-loving colleagues at the Institute of Arctic and Alpine Research; Mike Wireman of the EPA; Paul Tebbel at the National Audubon Society's Rowe Sanctuary; Stephen Hopp and his University of Arizona graduate students; Waverly Fitzgerald of School of the Seasons.com; and Elizabeth Howard, director of Journey North.

1

Frazer, James George, Sir. *The Golden Bough.* New York: Macmillan, 1922.

Newman, Mark F. N. et al. "Longitudinal assessment of neurocognitive function after coronary-artery bypass surgery." *The New England Journal of Medicine* v 344, n 6 (2001).

Parmesan, C. and Yohe, G. "A globally coherent fingerprint of climate change impacts across natural systems." *Nature* 421, 37–42 (2003).

Root, T. L. et al. "Fingerprints of global warming on wild animals and plants." *Nature* 421, 57–60 (2003).

Torrance, Robert M., ed. *Encompassing Nature, A Sourcebook.* Washington, D.C.: Counterpoint, 1998.

2

Mire, Pat. *Dance for a Chicken: The Cajun Mardi Gras.* Lafayette, LA: Pat Mire Films, 1993.

The Oxford English Dictionary. Oxford University Press, 1971.

Schwartz, Mark D., ed. *Phenology: An Integrative Science.* The Netherlands: Kluwer Academic Publishers, 2003.

3

Acuna-Soto, Rodolpho et al. "Megadrought and Megadeath in 16th Century Mexico." *Emerging Infectious Diseases Online, www.cdc.gov/eid* v 8, n 4 (April 2002).

Arizona-Sonora Desert Museum. *Migratory Pollinators Project.* Conservation and Science Department ASDM, 2003.

Cooper, Marc. "Take This Terrarium and Shove It." *Village Voice* (1991).

Schwartz, P. and Randall, D. "An Abrupt Climate Change Scenario and Its Implications for United States National Security." Emeryville, CA: Global Business Network, October 2003.

Siano, Brian. "The Skeptical Eye: Captain Future's Terrarium of Discipline." *The Humanist* v 52, n 2 (March 1992).

4

Millennium Ecosystem Assessment, 2005. *Ecosystems and Human Well-being: Synthesis.* Washington, D.C.: Island Press.

6

Wickman, Boyd E. "The Battle Against Bark Beetles in Crater Lake National Park: 1925–1934." *General Technical Report PNW-GTR-259,* U.S. Forest Service, June 1990.

8

Marshall, Robert. *Alaska Wilderness.* Berkeley: University of California Press, 1970.

———. *Arctic Village.* New York: H. Smith and Haas, 1933.

Acknowledgments

Special thanks to Michael Klemens, who understands spring and change more than anyone; Jim Malusa and his exuberant family—Sonya, Rudy, Rosita, and Abuela; Pat Mire, Paul and Donna Claus, Terry and Terri Smith, and Koy Chounlabout for sharing their time, knowledge, and friendship. To Doug Barasch and George Black at NRDC's *OnEarth* magazine and David Seidemann at Audubon. To Irene Schneider at *Condé Nast Traveler,* who has been both a great editor and friend.

Great thanks to the various doctors who took me apart and put me back together: Stephen Colvin, chairman of the Department of Cardio-thoracic Surgery at New York University Medical Center, who performed my transformative surgery, and his colleague, William J. Cole, cardiologist and literato. Harry Weinrauch, the most extraordinary physician, humanist, and reader I've ever known. Allen Collins, who would not allow me to dissuade myself from my path.

Finally, I thank Miles Groth, who allowed me the pleasure of Moby Dick's companionship for my trip; my agent, Dominick Abel, for his constancy; and Colin Harrison of Scribner—no one could ask for a more sentient editor. My children, Benjamin, Nathaniel, and Julia, to whom this book is dedicated. My mom gave me her support, love, and respect. Elaine Charnov gave me the substantial gifts of her humor, intelligence, and love.

Bruce Stutz, formerly features editor of *Audubon* magazine and editor-in-chief of *Natural History,* is a writer and editor whose articles on nature and travel have appeared in many newspapers and magazines including *The New York Times, The Washington Post, The Christian Science Monitor, The Atlantic Monthly, Islands, Travel & Leisure,* and *Condé Nast Traveler,* as well as *Scientific American* and *Discover.* He is also the author of a natural and cultural history of the Delaware River region titled *Natural Lives, Modern Times.* His interests in reporting on nature, science, environment, culture, and music have taken him around the world.